Algebra 2

Test Booklet

1·888·854·MATH(6284) - MathUSee.com
Sales@MathUSee.com

TESTS

Tests are primarily for evaluating a student's progress. If a student does well on the test, then he is ready to move on to the next lesson. If he does not do well, spend more time on that lesson and master it before moving to the new material. (The solutions for the tests are in the instruction manual.) Math is sequential and builds from concept to concept and from lesson to lesson. Master the material in each lesson before moving to the next topic.

The simplest way to solve a multiple-choice problem is to pretend there are no answers given and simply solve it, and then find your answer among those offered. But it is also a good idea to estimate the answer before doing any calculations and eliminate several of the possibilities. When the potential answers have been narrowed, then solve the problem and choose the correct answer.

You will find that this form of testing measures your reasoning abilities as well as your math knowledge. It also requires more than a cursory knowledge of the material being tested. Let me encourage you to look upon these tests, not merely as exams to conquer and pass, but also as opportunities to learn and stretch your knowledge about a particular topic.

The first 10 questions on each test cover the new concepts taught in that lesson. This is the material that must be mastered before moving on. The last five questions review material previously taught in Math-U-See *Algebra 1* and *Geometry*, or in previous lessons of *Algebra 2*. This booklet also contains three unit tests and a final exam, which are not written in multiple-choice format. They are designed to help you remember what you have already learned as you move through the course.

Before attempting the SAT or ACT tests, it is recommended that the student complete *Algebra I*, *Geometry*, and most of *Algebra 2*. We also suggest that you consider using prep books that are available for standardized testing.

Steve Demme

Circle your answer.

1. $(3^0)(3^{-2})(3^2) =$

    A. 3
    B. 1
    C. 9
    D. 0

2. $Y^8 \div Y^2 =$ $(Y \neq 0)$

    A. Y^6
    B. Y^4
    C. Y^{10}

    D. $\dfrac{1}{Y^6}$

3. $(3Q^2)^3 =$

    A. $3Q^6$
    B. $3Q^5$
    C. $9Q^6$
    D. $27Q^6$

4. $\dfrac{P^3 N^{-2}}{N^2 P^4} =$

    A. P^{-1}

    B. $\dfrac{P}{N^4}$

    C. $\dfrac{1}{N^4 P}$

    D. $N^4 P^{-1}$

5. If $3^{Y-1} = 81$, what is the value of Y?

    A. 5
    B. 4
    C. 3
    D. 2

6. If X is an integer, which of the following could **not** equal X^5?

    A. 0
    B. −1
    C. 16
    D. 32

7. The greatest common factor of $A^2 B^4 + B^3 A$ is:

    A. AB^3
    B. AB
    C. $A^2 B$
    D. $A^3 B$

8. Factoring out the greatest common factor from $P^2 Q + P^4 Q^2$ leaves:

    A. QP^2
    B. $Q + P$
    C. $P + Q^2$
    D. $1 + P^2 Q$

9. $(-2 + 4)^{-2} =$

    A. -4   B. -1/4   C. 1/4   D. 4

10. If $3^6 = 9^X$, X =

    A. 1   B. 2   C. 3   D. 4

Questions 1–10 on each test cover new concepts that must be mastered before moving on to the next lesson. Questions 11–15 cover concepts learned in previous courses or in previous lessons of *Algebra 2*. You may use these questions as a review tool.

11. If X + 2Y = 5 and X = 1/2 Y, then Y =

    A. 2   B. 1/2   C. 1   D. -2

12. In the rhombus shown, what is the value of *a*?

    A. 1   B. 2   C. 3   D. 4

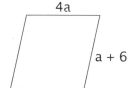

13. Three intersecting lines are shown. What is the value of b + a?

    A. 170   B. 160   C. 140   D. 70

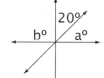

14. A recipe requires three eggs and seven cups of flour. If 15 eggs are used, how many cups of flour are needed?

    A. 5   B. 19   C. 25   D. 35

15. What is the slope of a line that passes through the origin and the point (-3, -2)?

    A. 2/3   B. -2/3   C. 3/2   D. -3/2

Circle your answer.

1. $\dfrac{5X+2}{5X} =$

   A. $\dfrac{2}{5X}$

   B. 3

   C. $1 + \dfrac{2}{5X}$

   D. 2

2. When adding two terms, you may:

   A. add one to either
   B. subtract one from either
   C. multiply either by one
   D. all of the above

3. $\dfrac{2X}{X+2} - \dfrac{3X}{X-2} =$

   A. $\dfrac{-2X(2X+5)}{(X+2)(X-2)}$

   B. $\dfrac{-X}{(X+2)(X-2)}$

   C. $\dfrac{4X^2+10X}{(X+2)(X-2)}$

   D. $\dfrac{-X^2-10X}{(X+2)(X-2)}$

4. $\dfrac{X^2+2X}{X} =$

   A. $X + 2$
   B. $4X$

   C. $\dfrac{X+2}{X}$

   D. cannot simplify

5. $\dfrac{9}{4X} - \dfrac{5}{4Y} =$

   A. $\dfrac{4X}{4XY}$

   B. $\dfrac{9Y-5X}{4XY}$

   C. $\dfrac{9Y^2-5X^2}{4XY}$

   D. $\dfrac{9Y-5X}{8XY}$

6. $\dfrac{A}{A} + A^0 =$ (assume $A \neq 0$)

   A. $\dfrac{A^2}{A}$

   B. $\dfrac{1}{A}$

   C. 1

   D. 2

7. $\dfrac{18AB-12A^2}{6A} =$

   A. $6A^2B$

   B. $3B - 2A$

   C. $\dfrac{6AB}{6A}$

   D. $3AB - 2A^2$

8. If two terms are to be combined, they must have the same:
   A. numerator
   B. coefficients
   C. denominator
   D. factors

9. $\dfrac{8Y}{X+1} + \dfrac{Y}{X-1} =$

   A. $\dfrac{16XY + 2Y}{(X+1)(X-1)}$

   B. $\dfrac{8Y + Y}{(X+1)(X-1)}$

   C. $\dfrac{9XY - 7Y}{(X+1)(X-1)}$

   D. $\dfrac{8Y^2}{(X+1)(X-1)}$

10. $\dfrac{2}{Y} + \dfrac{5}{3Y} - Y^{-1} =$

   A. $\dfrac{8}{3Y}$

   B. $\dfrac{2}{Y}$

   C. $\dfrac{14}{3Y}$

   D. $\dfrac{6}{Y}$

11. What number plus five equals two times the number?

   A. 2
   B. 3
   C. 4
   D. 5

12. $[(3^2)^{-3}]^{-1} =$

   A. 3^6
   B. 3^{-6}
   C. 3^{-2}
   D. 9^4

13. $-4^2 + 12 \div 4 - |6 - 8| =$

   A. -3
   B. -6
   C. -11
   D. -15

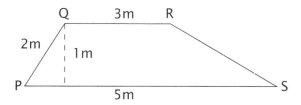

14. What is the area of PQRS?

   A. $10\ m^2$
   B. $18\ m^2$
   C. $14\ m^2$
   D. $4\ m^2$

15. Two lines that lie on the same plane are:

   A. congruent
   B. coplanar
   C. supplementary
   D. parallel

Circle your answer.   blue

1. 6,200 written in scientific notation is:

   A. 62×10^{-3}
   B. 6.2×10^{4}
   C. 6.2×10^{3}
   D. $.62 \times 10^{5}$

2. .268 written in scientific notation is:

   A. 2.68×10^{-2}
   B. 2.68×10^{-1}
   C. 2.68×10^{1}
   D. 26.8×10^{-1}

3. $.000073 \times .0054 =$

   A. 3.942×10^{7}
   B. 1.27×10^{-10}
   C. 3.942×10^{-7}
   D. 3.942×10^{-8}

4. $32,000,000 \div 16,000 =$

   A. 2.0×10^{3}
   B. 2.0×10^{-3}
   C. 1.6×10^{3}
   D. 2.0×10^{11}

5. $\dfrac{(2.3 \times 10^{-3})(4 \times 10^{4})}{2 \times 10^{-2}} =$

   A. 4.6×10^{-3}
   B. 4.6×10^{3}
   C. 9.2×10^{-3}
   D. 9.2×10^{3}

6. $2ab^{-1} + 3a^{-1}b - \dfrac{4b^{-1}}{a^{-1}} =$

   A. $\dfrac{b}{a}$
   B. $\dfrac{2a}{b} - \dfrac{b}{a}$
   C. $5ab - \dfrac{4a}{b}$
   D. $\dfrac{3b}{a} - \dfrac{2a}{b}$

7. $\dfrac{A}{6} + \dfrac{3}{A^{2}} =$

   A. $\dfrac{1}{2} + \dfrac{1}{A}$
   B. $\dfrac{A^{3} + 18}{6A^{2}}$
   C. $\dfrac{1}{2A}$
   D. $\dfrac{A + 3}{6A^{2}}$

8. $XXXY - YXXX + \dfrac{2}{Y^{-1}X^{-3}} =$

   A. $X^{3}Y + \dfrac{2}{X^{2}Y}$
   B. $4X^{3}Y$
   C. $2X^{3}Y$
   D. $\dfrac{2}{X^{3}Y} + 2X^{3}Y$

9. $3AABA^{-2} + 4AB - 6B =$

    A. $4AB + 3B$
    B. $4AB - 3B$
    C. $7AB - 6B$
    D. $3A^2B + 4AB - 6B$

10. $4X + \dfrac{3XY^2Y^{-1}}{Y^1} + 8XY =$

    A. $7X + 8XY$

    B. $4X + 11XY$

    C. $\dfrac{7X + 8XY}{Y}$

    D. $4X + 3XY^2 + 8XY$

11. If $\dfrac{1}{Y} - \dfrac{3Y}{Y} = 6$, then $Y =$

    A. 3
    B. 1/9
    C. 1/3
    D. 9

12. 5/8 written as a decimal is:

    A. 6.25
    B. .16
    C. 1.6
    D. .625

13. The least common multiple (LCM) of 3, 6, and 8 is:

    A. 24
    B. 18
    C. 48
    D. 144

14. What is the perimeter of the figure?

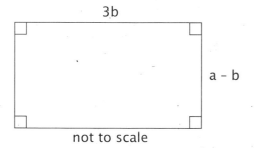

not to scale

    A. $2b + a$
    B. $6b - 2a$
    C. $3ab - 3b^2$
    D. $4b + 2a$

15. Which of the following is an obtuse angle?

    A. $23°$
    B. $230°$
    C. $102°$
    D. $90°$

Circle your answer.

1. $\left(2\sqrt{3}\right)\left(3\sqrt{2}\right) =$

   A. $6\sqrt{5}$
   B. $6\sqrt{6}$
   C. $8\sqrt{6}$
   D. $5\sqrt{5}$

2. $2\sqrt{3} + 3\sqrt{3} =$

   A. $6\sqrt{3}$
   B. $5\sqrt{6}$
   C. $5\sqrt{3}$
   D. $6\sqrt{6}$

3. $\dfrac{5}{\sqrt{3}} =$

   A. $\dfrac{5\sqrt{3}}{3}$

   B. $15\sqrt{3}$

   C. $\dfrac{5\sqrt{3}}{9}$

   D. $\dfrac{15}{\sqrt{3}}$

4. $2\sqrt{32} =$

   A. $12\sqrt{2}$
   B. $2\sqrt{2}$
   C. $8\sqrt{16}$
   D. $8\sqrt{2}$

5. $\sqrt{3}\left(\sqrt{18Y} + 3\sqrt{3}\right) =$

   A. $9\sqrt{6Y} + 27$
   B. $3\sqrt{6Y} + 9$
   C. $9\sqrt{3Y} + 6\sqrt{3}$
   D. $3\sqrt{6Y} + 27$

6. $\dfrac{\sqrt{75}}{\sqrt{5}} =$

   A. 15
   B. $3\sqrt{5}$
   C. $\sqrt{15}$
   D. $5\sqrt{3}$

7. $\dfrac{2}{\sqrt{5}} + \dfrac{5}{\sqrt{2}} =$

   A. $\dfrac{2\sqrt{5} + 5\sqrt{2}}{7}$

   B. $5\sqrt{2} + 2\sqrt{5}$

   C. $4\sqrt{5} + 2\sqrt{5}$

   D. $\dfrac{4\sqrt{5} + 25\sqrt{2}}{10}$

8. $\dfrac{2}{3}\sqrt{63Y^{12}} =$

   A. $2Y^6\sqrt{7}$

   B. $3Y^3\sqrt{63}$

   C. $\dfrac{2}{3}Y^4\sqrt{63}$

   D. $2Y^4$

9. $3\sqrt{\dfrac{9}{16}Y^4}$ =

    A. $9/4\ Y^2$
    B. $9/4\ Y$
    C. $4Y^2$
    D. $1/4\ Y^2$

10. $3\sqrt{50} - 2\sqrt{18}$ =

    A. $9\sqrt{2}$
    B. $15\sqrt{2} - 6\sqrt{3}$
    C. $15\sqrt{5} - 6\sqrt{2}$
    D. $24\sqrt{2}$

11. Two angles whose measures add up to 90° are called:

    A. right
    B. supplementary
    C. complementary
    D. vertical

12. Given $Y = 6X + 5$, the Y-intercept is:

    A. 1
    B. 5
    C. 6
    D. 11

13. What is the volume of a cylinder with a radius of *b* and a height of 6?

    A. $36b^2$
    B. $12\pi b$
    C. $12\pi b^2$
    D. $6\pi b^2$

14. Subtract 4 from Y.
    Multiply the difference by 6.
    Divide the product by 2.
    The result can be written as:

    A. $3Y - 12$

    B. $3Y - 2$

    C. $\dfrac{6Y - 4}{2}$

    D. $\dfrac{6(4 - Y)}{2}$

15. What is the fewest number of threes that can be multiplied together to yield a number greater than 60?

    A. 3
    B. 4
    C. 10
    D. 20

Circle your answer.

1. The factors of $2X^2 - 9X + 9$ are:

   A. $(2X + 3)(X - 3)$
   B. $(2X - 3)(X - 3)$
   C. $(2X - 9)(X - 3)$
   D. $(2X - 3)(2X - 3)$

2. The factors of $20 - 5X^2$ are:

   A. $(-5X - 2)(X + 2)$
   B. $5(X + 2)(X + 2)$
   C. $-5(X - 2)(X + 2)$
   D. $20(X - 2)(X + 2)$

3. The factors of $2X^3 - X^2 - 3X$ are:

   A. $X(X + 1)(2X - 3)$
   B. $(X + 1)(2X - 3)$
   C. $X(2X^2 - X + 3)$
   D. $(X - 1)(2X + 3)$

4. The factors of $Y^4 - 625$ are:

   A. $(Y - 5)(Y - 5)(Y^2 + 25)$
   B. $(Y^2 - 50)(Y^2 + 25)$
   C. $(Y - 5)(Y + 5)(Y^2 + 25)$
   D. $(Y^2 + 25)^2$

5. For $2X^2 + 4X = 6$, the values of X are:

   A. $-1, 3$
   B. $-3, 1$
   C. $2, -2$
   D. $6, -1$

6. For $-6X^2 = -27X + 12$, the values of X are:

   A. $1/2, 4$
   B. $2, -3$
   C. $1, 4$
   D. $-1/2, 2$

7. $\dfrac{4}{X+2} - \dfrac{2X}{2} =$

   A. $\dfrac{-2(X^2 + 2X - 4)}{X + 2}$
   B. $\dfrac{4 - 2X}{2(X + 2)}$
   C. $\dfrac{-(X^2 + 2X - 4)}{X + 2}$
   D. $\dfrac{2X}{X + 2}$

8. $\dfrac{3}{X+4} - \dfrac{2X}{4-X} + \dfrac{X^2}{X^2 - 16} =$

   A. $\dfrac{3X^2 + 11X - 12}{X^2 - 16}$
   B. $\dfrac{3X^2 - 11X}{X^2 - 16}$
   C. $\dfrac{X^2 - 2X + 3}{X^2 - 16}$
   D. $\dfrac{2X}{X + 2}$

9. $\dfrac{2 + \dfrac{6}{A}}{3 + \dfrac{12}{A-1}} =$

   A. $\dfrac{2A-1}{3A}$

   B. $\dfrac{2A-1}{A}$

   C. $\dfrac{A-1}{3A}$

   D. $\dfrac{2(A-1)}{3A}$

10. $\dfrac{\dfrac{X^2+9X+20}{X^3-9X}}{\dfrac{X^2+8X+16}{X^2+X-12}} =$

   A. $\dfrac{X(X+5)}{X}$

   B. $\dfrac{X+4}{X+3}$

   C. $\dfrac{X+5}{X(X+3)}$

   D. $\dfrac{(X+4)^2(X+5)}{X(X-3)^2}$

11. If $b + c = -3$, then $(b + c)^0 =$

   A. 0

   B. 1

   C. −1

   D. 3

12. Which of the following describes a line perpendicular to $Y = 1/3\ X + 5$?

   A. $Y = -1/3\ X + 5$

   B. $Y = 3X + 5$

   C. $Y = -3X + 5$

   D. $Y = 1/3\ X - 1/5$

13. There are seven coins, all either nickels or dimes. The value of the coins is 50 cents. How many nickels are there?

   A. 4

   B. 3

   C. 7

   D. 5

For #14 and #15, use these definitions for *n*.

$\left(\,n\,\right) = n^2 + 1$ $\overline{n} = n^2 - 2$

14. What is $\left(\,2\,\right) - \boxed{2}$?

   A. 0

   B. 3

   C. 4

   D. −1

15. What is $\left(\,B\,\right) + \boxed{B}$?

   A. $2B^2 - 3$

   B. $2B^2 + 2$

   C. $B^4 - 2$

   D. $2B^2 - 1$

Circle your answer.

Use fractional exponents to solve #6–10.

1. $1000^{2/3} =$

A. 1
B. 10
C. 100
D. 1000

2. $\left(\dfrac{81}{25}\right)^{\frac{1}{2}} =$

A. 5/9
B. 9/5
C. –5/9
D. –9/5

3. $\left(32^{3/5}\right)^2 =$

A. 8
B. 32
C. 64
D. 256

4. $\left(\dfrac{R^{1/3}}{2}\right)^2 =$

A. R
B. $R^{2/3}$
C. $2R^{3/2}$
D. $1/4R^{2/3}$

5. $\left(\dfrac{-1}{\sqrt{25}}\right)^{-3} =$

A. –5
B. 25
C. 125
D. –125

6. $\sqrt{\sqrt{81}} =$

A. 3
B. 6
C. 9
D. 12

7. $\sqrt[3]{B^6} =$

A. B^2
B. B^3
C. B^4
D. B^6

8. $\left(\sqrt[3]{27}\right)^4 =$

A. 3
B. 9
C. 27
D. 81

9. $\left(\sqrt[3]{64}\right)^{-2} =$

A. 1/4
B. 1/16
C. 4
D. 16

10. $\left(\dfrac{3}{\sqrt{4}}\right)^{-2} =$

A. 9/16
B. 4/9
C. 2/3
D. 9/4

11. The measures of the interior angles of a triangle add up to:

    A. 360°
    B. 180°
    C. 90°
    D. 270°

12. Which of the following is not true of the Cartesian coordinate system used for graphing?

    A. The X-axis is horizontal and the Y-axis is vertical.
    B. It can show negative and positive values.
    C. It can be used to show equations with two variables.
    D. It can be used to show equations with three variables.

13. What is the area of the unshaded part of the figure?

    A. 22 ft
    B. 24 ft^2
    C. 25 ft^2
    D. 30 ft^2

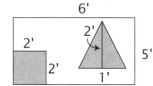

14. Mary got 30% of the test questions wrong. Sam missed twice as many. If there were 60 questions, how many correct answers did Sam have?

    A. 24
    B. 36
    C. 18
    D. 0

15. The sun is 93 million miles away. Alex walked three-tenths of a mile. How many times that distance would he have to walk in order to reach the sun?

    A. 3.1×10^6
    B. 2.8×10^{-6}
    C. 3.1×10^8
    D. 9.0×10^6

Circle your answer.

1. $\sqrt{-121}$ =

   A. 11
   B. 11i
   C. −11i
   D. 121i

2. $\sqrt{\dfrac{-81}{100}}$ =

   A. 9/10 i
   B. −9/10 i
   C. 81/100 i
   D. −81/100 i

3. $\sqrt{\dfrac{-16}{7}}$ =

   A. $\dfrac{4}{7}$

   B. $-\dfrac{4}{7}$

   C. $\dfrac{4\sqrt{7}}{7}i$

   D. $\dfrac{4}{7i}$

4. $\sqrt{-4} + \sqrt{-8}$ =

   A. 10i
   B. −2i + i$\sqrt{8}$
   C. 2i + 2i$\sqrt{2}$
   D. 2i − 4i$\sqrt{2}$

5. $\left(3\sqrt{-6}\right)\left(5\sqrt{-15}\right)$ =

   A. 15$\sqrt{90}$
   B. 1350i
   C. −150$\sqrt{3}$
   D. −45$\sqrt{10}$

6. $(i^3)^2$ =

   A. 1
   B. −1
   C. 0
   D. i

7. $\sqrt{81} - \sqrt{-4}$ =

   A. 9 − 2i
   B. 9 + 2i
   C. 7
   D. −7

8. (7i)(−3i) =

   A. −21
   B. 21
   C. $\sqrt{21}$
   D. 21i

9. $\left(2\sqrt{-4}\right)\left(5\sqrt{-9}\right)$ =

   A. 10$\sqrt{13}$
   B. 60i
   C. −60
   D. 60

10. $[(3i)(4i)]^2$

    A. 12   B. –12   C. 144   D. –144

11. Which statement about line *a* is **not** true?

    A. The slope is 2.
    B. The slope is –2.
    C. The Y-intercept is 2.
    D. The X-intercept is 1.

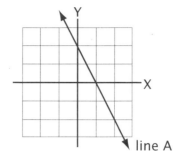

line A

12. The equation for line *a* is:

    A. Y = 2X + 2
    B. Y = 1/2 X + 2
    C. X = 2Y + 1
    D. Y = –2X + 2

13. The area of a rectangle is $6X^2$. If one dimension is 2X, what is the length of the other dimension?

    A. 4X   B. $3X^2$ C. 3X   D. 12X

14. The radius of a circle is:

    A. one-half the diameter
    B. twice the diameter
    C. π times the circumference
    D. one-half the circumference

15. Megan is standing on a point that is 40°16' N. Sarah is at 39°20' N and on the same line of longitude as Megan. In what direction must Megan travel in order to join Sarah?

    A. north   B. south   C. east   D. west

Circle your answer.

1. The conjugate of X – 2i is:

   A. X
   B. X + 2i
   C. X – 2
   D. 2 – X

3. The conjugate of 7 – i is:

   A. i – 7
   B. 7 + i
   C. 7i
   D. 7 – i

2. The conjugate of $3+\sqrt{2}$ is:

   A. 5
   B. $3+\sqrt{2}$
   C. $3-\sqrt{2}$
   D. 1

4. The conjugate of $2+\sqrt{A}$ is:

   A. $2-\sqrt{A}$
   B. $2+\sqrt{A}$
   C. 2A
   D. $\sqrt{A}-2$

For #5–10, simplify or put in standard form.

5. $\dfrac{Y}{4-3i}=$

   A. $\dfrac{4Y+3Yi}{7}$

   B. $\dfrac{7}{4Y+3Yi}$

   C. $\dfrac{25}{4Y+3Yi}$

   D. $\dfrac{4Y+3Yi}{25}$

7. $\dfrac{-6}{9-3\sqrt{3}}=$

   A. –2

   B. $\dfrac{-3-\sqrt{3}}{3}$

   C. $\dfrac{6+2\sqrt{3}}{8}$

   D. $\dfrac{2\sqrt{3}}{6}$

6. $\dfrac{5Q}{2+\sqrt{7}}=$

   A. $\dfrac{5\sqrt{7}Q-10Q}{-45}$

   B. $\dfrac{10Q+5\sqrt{7}Q}{3}$

   C. $\dfrac{10Q-5\sqrt{7}Q}{-3}$

   D. $\dfrac{10Q+5\sqrt{7}Q}{-11}$

8. $\dfrac{2X+1}{i}=$

   A. $\dfrac{2X+1}{i}$

   B. $\dfrac{i}{2X+1}$

   C. –2Xi – i

   D. 2Xi + i

9. $\dfrac{i}{1+\sqrt{2}} =$

   A. -1

   B. $\sqrt{2}$

   C. $i - i\sqrt{2}$

   D. $i\sqrt{2} - i$

10. $\dfrac{3}{2 - \sqrt{Y}} =$

   A. $\dfrac{6 - 3\sqrt{Y}}{4 - Y}$

   B. $\dfrac{6 - 3\sqrt{Y}}{Y - 4}$

   C. $\dfrac{6 + 3\sqrt{Y}}{Y - 4}$

   D. $\dfrac{6 + 3\sqrt{Y}}{4 - Y}$

11. Lines a and b are parallel. Line c is a transversal. Which pair of angles are alternate exterior angles?

   A. 1 and 2
   B. 1 and 4
   C. 2 and 7
   D. 2 and 8

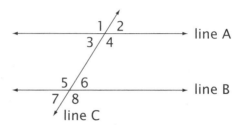

12. Pairs of alternate exterior angles are always:

   A. congruent
   B. unequal
   C. complementary
   D. obtuse

13. An engineer needs values for X and Y that satisfy both of the following equations: X – Y = 3 and 3X – Y = 13. The needed values are:

   A. (8, 2)
   B. (–5, 2)
   C. (8, 5)
   D. (5, 2)

14. Ariana drew lines to make four triangles inside a regular hexagon. Use the drawing and the fact that there are 180º in a triangle to find the sum of the interior angles of a hexagon.

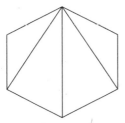

    A. 120º

    B. 720º

    C. 1080º

    D. 360º

15. How many degrees are there in each interior angle of a regular hexagon?

    A. 60º

    B. 120º

    C. 180º

    D. 100º

For #1 and #2, find the binomial roots of the trinomial.

1. $X^2 + 10X + 25$

    A. $(X + 5)(X - 5)$
    B. $(X + 5)^3$
    C. $(X + 5)^2$
    D. $(2X + 5)^2$

2. $4A^2 + 16A + 16$

    A. $(2A + 4)^2$
    B. $(2A + 4)(2A - 4)$
    C. $(4A + 8)^2$
    D. $(2A + 4)^3$

For #3–6, expand the binomial.

3. $(X + 4)^2 =$

    A. $X^2 + 2X + 4$
    B. $X^2 + 8X + 16$
    C. $2X^2 + 8X + 8$
    D. $X^2 + 8X - 16$

4. $(2X - 2)^2 =$

    A. $4X^2 - 8X + 4$
    B. $4X^2 - 8X - 4$
    C. $X^2 - 4X + 4$
    D. $2X^2 - 4X + 4$

5. $(X + 2)^3 =$

    A. $X^2 + 4X + 4$
    B. $3X^2 + 6X + 6$
    C. $X^3 + 3X^2 + 6X + 8$
    D. $X^3 + 6X^2 + 12X + 8$

6. $(A - B)^3 =$

    A. $A^3 + 3A^2B + 3AB^2 + B^3$
    B. $A^3 - 3A^2B + 3AB^2 - B^3$
    C. $2A^3 - 6A^2B + 6AB^2 - B^3$
    D. $A^3 - 3AB^2 + 3A^2B - B^3$

7. The numerals in Pascal's triangle represent:

    A. the exponents of a polynomial
    B. the coefficients of a polynomial
    C. the possible values of X
    D. the values of all the variables

8. The first term of $(Y - 10)^3$ is:

    A. Y^2
    B. $10Y^3$
    C. Y^3
    D. $-10Y^3$

9. The second term of $(3X - 1)^3$ is:

    A. $-27X^2$
    B. $9X^2$
    C. $27X^2$
    D. -1

10. The last term of $(B + 1/2)^3$ is:

    A. 1/2
    B. 1/4
    C. 1/8 B^3
    D. 1/8

11. The area of a rectangle is 864 in^2. What is the area of the same rectangle expressed in ft^2?

    A. 6 ft^2
    B. 144 ft^2
    C. 72 ft^2
    D. 288 ft^2

12. In the triangle shown, what is the measure of angle ABC?

    A. 115°
    B. 65°
    C. 39°
    D. 51°

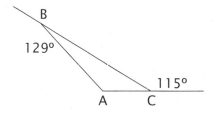

13. In the triangle shown, what is the measure of angle A?

    A. 26°
    B. 65°
    C. 64°
    D. cannot be determined

14. Our number system is expressed in:

    A. base 12
    B. base 2
    C. base 10
    D. base 1

15. In base 2, the number 3 would be written as:

    A. 111_2
    B. 11_2
    C. 100_2
    D. 101_2

Circle your answer.

1. When a binomial is raised to any power and expanded, the exponents of the A term:

    A. begin with the given power and decrease by one for each term
    B. begin with zero and increase by one for each term
    C. begin with one and increase by one for each term
    D. remain the same for each term

2. When a binomial is raised to any power and expanded, the exponents of the B term:

    A. begin with the given power and decrease by one for each term
    B. begin with zero and increase by one for each term
    C. begin with one and increase by one for each term
    D. remain the same for each term

3. When a binomial is raised to a given power and expanded, the number of terms is:

    A. the power plus one
    B. the power minus one
    C. the same as the power
    D. none of the above

4. If $(A + B)^7$ is expanded, the exponents of A and B in each term will add to:

    A. 8
    B. 6
    C. 7
    D. a different sum in each term

5. The binomial theorem:

    A. is used to solve binomial equations
    B. has no relationship to Pascal's triangle
    C. applies only to expressions raised to the second power
    D. can be used to predict any term of a binomial raised to a given power

6. If $(3X + 4)^6$ is expanded, how many terms will it have?

    A. 5
    B. 7
    C. 6
    D. 12

7. What is the third term in $(X + Y)^6$?

   A. $15X^4Y^2$
   B. $12X^4Y^2$
   C. $15X^2Y^4$
   D. $15X^3Y^3$

8. What is the fourth term in $(2A + B)^5$?

   A. $20A^2B^2$
   B. $40A^3B^2$
   C. $40A^2B^3$
   D. $40A^2B^2$

9. What is the fifth term in $(X - 1/2)^4$?

   A. no fifth term
   B. $-1/8$
   C. $-1/16$
   D. $1/16$

10. What is the second term in $(X + 2Y)^7$?

    A. $14XY^6$
    B. $14X^7Y^0$
    C. $128XY^7$
    D. $14X^6Y$

11. The legs of a right triangle have lengths of A and 2A. What is the length of the hypotenuse?

    A. $A\sqrt{5}$
    B. $5A^2$
    C. $3A$
    D. $\sqrt{3A}$

12. What is the area of the regular polygon shown below?

    A. $12\sqrt{3}$ in^2
    B. $48\sqrt{3}$ in^2
    C. $24\sqrt{3}$ in^2
    D. 72 in^2

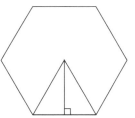

Line from center perpendicular to side = $2\sqrt{3}$ ".
Length of each side = 4".

13. The length of one leg of a 45°–45°–90° triangle is b. What must b be multiplied by to find the length of the hypotenuse?

    A. b
    B. $b\sqrt{2}$
    C. $\sqrt{2}$
    D. 2

14. Ten pounds is about equal to:

    A. 20 kilograms
    B. 5 kilograms
    C. 10 kilograms
    D. 1 kilogram

15. A rectangle measures X ft by Y ft. If the measurements are changed to 2X by 2Y ft, the area will be:

    A. one-half as great
    B. twice as great
    C. four times as great
    D. none of the above

Simplify.

1. $(2^0)(2^{-3})(2^3) =$

2. $X^9 \div X^3 =$

3. $\left(4Q^3\right)^2 =$

4. $\dfrac{X^3 Y^{-2}}{Y^2 X^4} =$

Add or Subtract.

5. $\dfrac{3}{4A} - \dfrac{8}{4B} =$

6. $\dfrac{R}{R} + R^0 =$

7. $\dfrac{2Y}{X+Y} + \dfrac{Y}{X-Y} =$

8. $4R^6 TR^{-2} + 5R^4 T - 2T =$

Solve using scientific notation.

9. $(.0056)(.034) =$

10. $(45,500)(21,000,000) =$

11. $(32,000) \div (.00016) =$

12. $\dfrac{(.00023)(160)}{.002} =$

Simplify.

13. $\left(4\sqrt{5}\right)\left(5\sqrt{3}\right) =$

14. $5\sqrt{6} + 2\sqrt{6} =$

15. $\dfrac{6}{\sqrt{2}} + \dfrac{1}{\sqrt{3}} =$

16. $\sqrt{36X^4} =$

17. $(\dfrac{16}{25})^{\frac{1}{2}} =$

18. $\sqrt{\sqrt{16}} =$

Find the factors.

19. $3X^2 + 17X + 10$

20. $3X^2 - 9X + 6$

21. $X^4 - 1$

22. $2X^2 + 3X - 2$

Solve for the unknown.

23. $X^2 - 10X = -18 - X$

24. $2X^2 + 2X + 14 = 32 + 2X$

25. $2X + 15 = X^2$

26. $X^3 = 16X$

Simplify, and combine like terms where possible.

27. $\sqrt{-144}$ =

28. $\sqrt{-8} + \sqrt{-4}$ =

29. $\left(4\sqrt{-5}\right)\left(2\sqrt{-6}\right)$ =

30. $(i^3)^2$ =

Simplify, or put in standard form.

31. $\dfrac{X}{8 + 2i}$ =

32. $\dfrac{2}{1 + \sqrt{2}}$

Answer the questions.

33. What is the third term in $(X + Y)^5$?

34. What is the second term in $(A + B)^4$?

35. What is the fourth term in $(D - 1/2)^6$?

For #1–3, complete the square by finding the last term.

1. $X^2 + 16X +$ _____

   A. 8
   B. 16
   C. 256
   D. 64

2. $X^2 - 7X +$ _____

   A. 49/4
   B. 49
   C. 49/2
   D. -49/4

3. $X^2 + 3/4\ X +$ _____

   A. 9/64
   B. 9/16
   C. 9/4
   D. 6/8

For #4–6, complete the square by finding the middle term.

4. $X^2 -$ _____ $+ 81$

   A. 9X
   B. 18X
   C. 3 1/2 X
   D. 18

5. $X^2 +$ _____ $+ 144$

   A. 12X
   B. 6X
   C. 24X
   D. 12

6. $X^2 +$ _____ $+ 16/25$

   A. 8/5 X
   B. 4/10 X
   C. 4/5 X
   D. 2/5 X

7. The square root of $X^2 - 14X + 49$ is:

   A. X – 14
   B. 7X
   C. X + 7
   D. X – 7

For #8–10, solve for X by completing the square.

8. $X^2 + 8X - 4 = 0$

   A. $X = -4 \pm 2\sqrt{3}$
   B. $X = -4 \pm 2\sqrt{5}$
   C. $X = 4 \pm 2\sqrt{3}$
   D. $X = -8 \pm 2\sqrt{5}$

9. $X^2 - 10X + 3 = 0$

   A. $X = -5 \pm \sqrt{22}$
   B. $X = 10 \pm \sqrt{22}$
   C. $X = 5 \pm 2\sqrt{7}$
   D. $X = 5 \pm \sqrt{22}$

10. $X^2 + 4X + 8 = 0$

    A. $X = -2 \pm 2\sqrt{3}$
    B. X = 0, –4
    C. $X = -2 \pm 2i$
    D. $X = 2 \pm 2i$

11. The hypotenuse of a 30°-60°-90° triangle is 8 units and one leg is 4 units. What is the length of the other leg?

    A. $4\sqrt{2}$ units
    B. 4 units
    C. 8 units
    D. $4\sqrt{3}$ units

12. Twelve liters is about the same as:

    A. 12 qt
    B. 6 qt
    C. 24 qt
    D. 8 qt

13. A triangle with no congruent sides is said to be:

    A. isosceles
    B. scalene
    C. acute
    D. equilateral

14. Which of the following postulates is **not** true?

    A. A midpoint divides a line segment into two congruent segments.
    B. Two angles whose measures add up to 180° are supplementary.
    C. The sum of the measures of the interior angles of a triangle are 360°.
    D. If A = B and B = C, then A = C.

15. Two lines that intersect and form a right angle are:

    A. parallel
    B. perpendicular
    C. congruent
    D. reflexive

Circle your answer.

1. Which of the following cannot be solved using the quadratic equation?

   A. $X^2 - 64 = 0$
   B. $X^3 + 3Y + 1 = 0$
   C. $4A^2 + 8A = 16$
   D. $Y^2 = 2Y + 4$

2. The part of the quadratic formula written under the radical is:

   A. $B^2 + 4AC$
   B. $B^2 - 4AC$
   C. $-B^2 \pm 4AC$
   D. $A^2 + 4BC$

3. All quadratic equations can be solved by:

   A. factoring
   B. both factoring and the quadratic formula
   C. the quadratic formula
   D. none of the above

4. In order to find values of A, B, and C in the quadratic formula, an equation should be in the form:

   A. $AX^2 = BX + C$
   B. $X^2 + AX = B - C$
   C. $AX^2 + BX + C = 0$
   D. $AX^2 + BX = -C$

5. The solution to $7X^2 + 2X - 1 = 0$ can be written as:

   A. $X = \dfrac{-2 \pm \sqrt{2^2 - (4)(7)(-1)}}{2(7)}$

   B. $X = \dfrac{2 \pm \sqrt{2^2 - (4)(7)(-1)}}{2(7)}$

   C. $X = \dfrac{-2 \pm \sqrt{2^2 + (4)(7)(-1)}}{2(7)}$

   D. $X = \dfrac{-2 \pm \sqrt{(-2)^2 - (4)(7)(-1)}}{2}$

For #6–10, solve using the best method.

6. $X^2 - 36 = 0$

   A. $X = 6, -6$
   B. $X = 4, 9$
   C. $X = 0, 6$
   D. $X = \pm 9$

7. $X^2 + 3 = -3X$

   A. $X = \dfrac{-3 \pm \sqrt{3}}{2}$

   B. $X = \dfrac{-3 \pm i\sqrt{3}}{6}$

   C. $X = \dfrac{3 \pm i\sqrt{3}}{2}$

   D. $X = \dfrac{-3 \pm i\sqrt{3}}{2}$

8. $5X^2 = -2X + 1$

   A. $X = \dfrac{-1 \pm \sqrt{5}}{5}$

   B. $X = \dfrac{-1 \pm \sqrt{6}}{5}$

   C. $X = \dfrac{1 \pm 2\sqrt{6}}{5}$

   D. $X = \dfrac{1 \pm \sqrt{5}}{5}$

9. $4X^2 + 20X = -25$

   A. $X = \pm 5/2$
   B. $X = 4, 5$
   C. $X = 5/2$
   D. $X = -5/2$

10. $4X^2 + 4X - 10 = 0$

   A. $X = \dfrac{-1 \pm i\sqrt{11}}{2}$

   B. $X = i, -2i$

   C. $X = \dfrac{-1 \pm \sqrt{11}}{2}$

   D. $X = \dfrac{-1 \pm 3i}{2}$

11. $\triangle ABC$ is congruent to $\triangle EDC$. $\overline{AB}$ corresponds to:

   A. $\overline{BA}$
   B. $\overline{AC}$
   C. $\overline{ED}$
   D. $\overline{BC}$

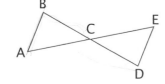

12. A quadrilateral with only one pair of parallel sides is a:

   A. rhombus
   B. trapezoid
   C. parallelogram
   D. regular polygon

13. Two sides of triangle A are congruent to the corresponding sides of triangle B. The angle formed by the corresponding sides is 25° in both triangles. What postulate may be used to prove triangles A and B congruent?

   A. SSS
   B. SSA
   C. SAS
   D. cannot be proved congruent

14. Each angle of triangle ABC is congruent to the corresponding angle of triangle DEF. What postulate may be used to prove $\triangle ABC$ and $\triangle DEF$ congruent?

   A. SSS
   B. AAA
   C. SAS
   D. cannot be proved congruent

15. Five yards are a little less than:

   A. 5 meters
   B. 10 meters
   C. 2 meters
   D. 6 meters

Circle your answer.

1. If the discriminant is equal to zero, the roots are:

   A. real, rational, unequal
   B. real, rational, equal
   C. real, irrational, unequal
   D. imaginary

2. If the discriminant is a perfect square, the roots are:

   A. real, rational, unequal
   B. real, rational, equal
   C. real, irrational, unequal
   D. imaginary

3. If the discriminant is negative, the roots are:

   A. real, rational, unequal
   B. real, rational, equal
   C. real, irrational, unequal
   D. imaginary

4. If the discriminant is greater than zero and not a perfect square, the roots are:

   A. real, rational, unequal
   B. real, rational, equal
   C. real, irrational, unequal
   D. imaginary

5. The discriminant of $X^2 + X = X + 9$ is:

   A. 40
   B. –36
   C. 36
   D. 9

6. The discriminant of $X^2 + 5 = 2X$ is:

   A. 17
   B. 16
   C. 24
   D. –16

7. The discriminant of $X^2 + 9 = -6X$ is:

   A. 33
   B. 0
   C. –33
   D. 72

For #8–10, solve for X.

8. $X^2 - 32 = -4X$

   A. X = –8, 4
   B. X = –2 ± i
   C. $X = \dfrac{-4 \pm i}{2}$
   D. X = –4, 8

9. $X^2 + 3X - 6 = 0$

   A. $X = \dfrac{-3 \pm \sqrt{33}}{2}$

   B. $X = \dfrac{-3 \pm i\sqrt{15}}{2}$

   C. $X = \dfrac{-3 \pm \sqrt{15}}{2}$

   D. $X = -2, 3$

10. $X^2 - 5X = -8$

   A. $X = \dfrac{-5 \pm \sqrt{7}}{2}$

   B. $X = \dfrac{5 \pm \sqrt{7}}{2}$

   C. $X = \dfrac{5 \pm i\sqrt{7}}{2}$

   D. $X = \dfrac{-5 \pm \sqrt{7}}{2}$

11. Right triangles can be proved congruent with less information than other triangles because:

   A. all right triangles are congruent
   B. all right triangles are equilateral
   C. their angles add up to 180°
   D. one set of congruent angles is given by definition

12. $62{,}000 \times .75 =$
    (Ignore significant digits for this problem.)

   A. 4.65×10^3
   B. 4.65×10^4
   C. 4.65×10^5
   D. 4.65×10^2

13. Two polygons that are not identical in size, but have the same proportions, are said to be:

   A. similar
   B. congruent
   C. equal
   D. rational

14. The two Rs on the graph on the next page are an example of:

   A. translation
   B. dilation
   C. rotation
   D. reflection

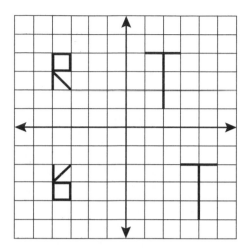

15. The two Ts on the graph are an example of:

    A. translation

    B. dilation

    C. rotation

    D. reflection

Circle your answer.

1. The original price of the table was $250. Jane bought it on sale for $200. What percent of the original price did she save?

   A. 10%
   B. 80%
   C. 20%
   D. 25%

2. The wholesale price of the skirt is $12 and the retail price is $24. What percent of the wholesale price is the markup?

   A. 100%
   B. 50%
   C. 12%
   D. 5%

3. Emily's wages were increased from $7.25 an hour to $7.83 an hour. What percent of her original rate was the raise?

   A. 80%
   B. 7%
   C. 8%
   D. 93%

4. Jason sold his house for $100,000. That was a gain of 25% over what he paid for the house. What did he pay for the house?

   A. $25,000
   B. $75,000
   C. $45,000
   D. $80,000

5. The cost of the food at the restaurant was $38.95. The tax was 5% and the tip was 20%. What was the final cost of the meal? (round to the nearest cent)

   A. $48.69
   B. $29.21
   C. $54.53
   D. $49.08

6. In the spring, boots were advertised for 45% off the marked price. What would Brad pay for a pair of boots marked $75?

   A. $41.25
   B. $33.75
   C. $108.75
   D. $30.00

7. A merchant planned on a markup of 40% over wholesale. What price would he put on an item with a wholesale cost of $32?

    A. $12.80
    B. $44.80
    C. $72.00
    D. $33.28

For #8–10, use these atomic weights. Round your answers to the nearest percent.

    hydrogen (H) = 1
    carbon (C) = 12
    oxygen (O) = 16
    sodium (Na) = 23
    chlorine (Cl) = 35

8. What is the percentage of chlorine in NaCl?

    A. 40%
    B. 6%
    C. 4%
    D. 60%

9. What is the percentage of oxygen in CO_2?

    A. 57%
    B. 27%
    C. 43%
    D. 73%

10. What is the percentage of sodium in NaOH?

    A. 28%
    B. 40%
    C. 58%
    D. 57%

11. $(9^{-2}) =$

    A. 81

    B. $\dfrac{1}{81}$

    C. -81

    D. -18

12. $\dfrac{Y}{X} + \dfrac{4Y}{X+2} =$

    A. $\dfrac{5YX + 2Y}{X(X+2)}$

    B. $\dfrac{YX + 2}{X^2 + 2X}$

    C. $\dfrac{5Y}{X(X+2)}$

    D. $\dfrac{7Y}{X+2}$

13. The shaded section represents 15% of the circle, which has a radius of three. To the nearest hundredth, what is the area of the remaining (unshaded) part of the circle?

    A. 28.26
    B. 24.02
    C. 4.24
    D. 16.01

14. The formula for the discriminant is based on:

   A. Pascal's triangle
   B. the binomial theorem
   C. the quadratic formula
   D. conjugate numbers

15. The simplest form of the quantity two times the square root of five, multiplied by the quantity five times the square root of twelve is:

   A. $40\sqrt{15}$
   B. $10\sqrt{60}$
   C. 300
   D. $20\sqrt{15}$

Circle your answer.

1. Solve for A: ABC = D

    A. A = D − BC

    B. A = BCD

    C. $A = \dfrac{BC}{D}$

    D. $A = \dfrac{D}{BC}$

2. Solve for B: $\dfrac{YZ}{B} = \dfrac{A}{X}$

    A. B = AXYZ

    B. $B = XY^2 − A$

    C. $B = \dfrac{XYZ}{A}$

    D. $B = \dfrac{A}{XYZ}$

3. Solve for Q: $\dfrac{Q}{P} − R = 0$

    A. Q = RP

    B. Q = −RP

    C. $Q = \dfrac{R}{P}$

    D. $Q = \dfrac{P}{R}$

4. Solve for X: X(Y − Z) + D = 4

    A. $X = \dfrac{4+D}{X−Z}$

    B. X = 4 − Y + Z − D

    C. $X = \dfrac{Y−Z}{4−D}$

    D. $X = \dfrac{4−D}{Y−Z}$

5. Solve for B: $\dfrac{1}{B} = \dfrac{1}{C}$

    A. B = −C

    B. B = C

    C. $B = \dfrac{1}{C}$

    D. B = 2C

6. Solve for Y: $\dfrac{X}{YZ} = \dfrac{S}{T}$

    A. $Y = \dfrac{SZ}{TX}$

    B. $Y = \dfrac{TX}{SZ}$

    C. Y = TX − SZ

    D. Y = STXZ

7. Solve for A: $\dfrac{RS}{T} = \dfrac{B}{A}$

    A. A = BT − RS

    B. $A = \dfrac{RS}{BT}$

    C. $A = \dfrac{BT}{RS}$

    D. $A = \dfrac{RSB}{T}$

8. Solve for Y: X − Z = Y + 5

    A. Y = X − Z − 5

    B. Y = 5 − X − Z

    C. Y = 5 + X − Z

    D. $Y = \dfrac{X - Z}{5}$

9. Solve for B: A(B + C) − D = X

    A. $B = \dfrac{X + D}{AC}$

    B. $B = \dfrac{X + D - C}{A}$

    C. $B = \dfrac{X - D}{A} - C$

    D. $B = \dfrac{X + D}{A} - C$

10. Solve for X: −X + Y − 4 = A + B

    A. X = A + B − Y + 4

    B. X = −A − B + Y − 4

    C. $\dfrac{A + 3}{y - 4}$

    D. $X = \dfrac{4(A + B)}{y}$

11. What is the standard form of $\dfrac{3}{2\sqrt{8}}$?

    A. $\dfrac{3}{8}$

    B. $\dfrac{3\sqrt{2}}{8}$

    C. $\dfrac{3}{2\sqrt{8}}$

    D. $\dfrac{3}{4\sqrt{2}}$

12. If $X^2 + BX + C$ can be factored, which of the following statements are true?

    A. The factors of C are the addends of B.

    B. The factors of B are the addends of C.

    C. The value of B is two times the value of C.

    D. B and C have no predictable relationship.

13. $6 + (-8) - 4 + 3 =$

    A. 1

    B. −3

    C. −9

    D. 13

14. A path 10 ft long forms the diameter of a circular flower bed. How long is the path around the outside of the flower bed?

    A. a little more than three times as long
    B. twice as long
    C. a little less than three times as long
    D. as long as the radius squared

15. Fraction A divided by fraction B is the same as:

    A. fraction B divided by fraction A
    B. fraction A divided by the reciprocal of fraction B
    C. fraction A times the reciprocal of fraction B
    D. fraction A times fraction B

Circle your answer.

1. The team played 32 games this season. The ratio of wins to losses was three to five. Which ratio cannot be derived from the given information?

   A. $\dfrac{\text{wins}}{\text{losses}} = \dfrac{3}{5}$

   B. $\dfrac{\text{losses}}{\text{total}} = \dfrac{5}{8}$

   C. $\dfrac{\text{wins}}{\text{total}} = \dfrac{3}{8}$

   D. $\dfrac{\text{losses}}{\text{total}} = \dfrac{5}{15}$

2. Tom made 52 attempts to hit the target. His ratio of success to failure was three to one. Which of the following ratios is true?

   A. $\dfrac{\text{successes}}{52} = \dfrac{3}{1}$

   B. $\dfrac{\text{successes}}{52} = \dfrac{3}{4}$

   C. $\dfrac{\text{successes}}{52} = \dfrac{1}{3}$

   D. $\dfrac{\text{failures}}{52} = \dfrac{3}{4}$

3. The ratio of red to purple is two to seven. If 21 are purple, how many are red?

   A. 21
   B. 9
   C. 5
   D. 6

4. A total of 243 mosquitoes filled the room, but only females bite. If the ratio of males to females is four to five, how many of the mosquitoes could bite?

   A. 135

   B. 108

   C. 304

   D. 27

5. 540 people visited the gift shop. If two visitors bought something for every three who only looked, how many people only looked?

   A. 324

   B. 108

   C. 216

   D. 540

6. The ratio of rainy days to sunny days was four to five. If there were 100 sunny days, how many were rainy?

   A. 44
   B. 20
   C. 80
   D. 91

For #7–10, use these atomic weights.

hydrogen (H) = 1, carbon (C) = 12,
nitrogen (N) = 14, oxygen (O) = 16,
sulfur (S) = 32, potassium (K) = 39

7. What is the mass of sulfur in 798
grams of CS_2?

    A. 1596 g
    B. 672 g
    C. 126 g
    D. 150 g

8. What is the mass of hydrogen in
720 grams of water (H_2O)?

    A. 40 g
    B. 80 g
    C. 680 g
    D. 18 g

9. What is the mass of oxygen in
1440 grams of H_2O?

    A. 1355 g
    B. 1620 g
    C. 1280 g
    D. 160 g

10. Which set of ratios should be used
to find the mass of potassium (K)
in 455 grams of KCN?

    A. $\dfrac{K}{455} = \dfrac{39}{65}$

    B. $\dfrac{K}{39} = \dfrac{12}{65}$

    C. $\dfrac{K}{455} = \dfrac{65}{39}$

    D. $\dfrac{39}{KCN} = \dfrac{14}{455}$

11. $\sqrt{\sqrt[3]{X}}$ is the same as:

    A. $X^{3/2}$
    B. X^6
    C. $X^{2/3}$
    D. $X^{1/6}$

12. The square root of –1 is:

    A. real
    B. negative
    C. imaginary
    D. complex

13. Complete the square by finding the
last term for $X^2 - 18X +$ _____.

    A. 81
    B. –81
    C. 18
    D. 36

14. In the equation $X^2 - X + 2X = 6$, $X = Y - 1$. What is one possible value of Y?

    A. 2
    B. 3
    C. −3
    D. 0

15. A tube with a diameter of 2 cm runs through the center of a container that is 10 x 10 x 10 cm. How many cubic cm of space are left in the container?

    A. 874.4 cm^3
    B. 996.9 cm^3
    C. 68.6 cm^3
    D. 968.6 cm^3

Circle your answer.

1. The key to unit multipliers is:

   A. memorizing a formula
   B. adding one to each unit
   C. always dividing the larger unit by the smaller
   D. multiplying by a fraction equal to one

2. How many unit multipliers are needed to change five square feet to square inches?

   A. 1
   B. 2
   C. 5
   D. 12

3. 80 ounces = _____ pounds

   A. 6.7
   B. 960
   C. 1280
   D. 5

4. 6 yards = _____ feet

   A. 18
   B. 36
   C. 2
   D. 72

5. Change 360 cm^3 into cubic meters.

   A. 3.60 m^3
   B. .00036 m^3
   C. 3,600 m^3
   D. 36,000 m^3

6. How many square feet are there in three square miles?

   A. 15,840 ft^2
   B. 1,760 ft^2
   C. 83,635,200 ft^2
   D. 47,520 ft^2

For # 7–9, use these metric/imperial conversions as needed.

   1 mi = 1.6 km
   1 km = .62 mi
   1 oz = 28 g
   1 g = .035 oz
   1 qt = .95 liters
   1 liter = 1.06 qt

7. How many miles are there in nine kilometers?

   A. 14.4 mi
   B. 14.52 mi
   C. 18 mi
   D. 5.58 mi

8. How many grams are there in 56 ounces?

   A. 2 g
   B. 1.96 g
   C. 1,568 g
   D. 3.5 g

9. Change 10 quarts to liters.

   A. 9.5 liters
   B. 40 liters
   C. 10.6 liters
   D. 2 liters

10. Devan invented a new system of measurement. He named the units after colors. There are three reds in one blue and five blues in one green. How many reds are there in two square greens?

   A. 225 reds^2
   B. 450 reds^2
   C. 25 reds^2
   D. 30 reds^2

11. The conjugate of $4 + \sqrt{10}$ is:

   A. $4 - \sqrt{10}$

   B. $4 - 10$

   C. $4 + \sqrt{10}$

   D. $\dfrac{1}{4 + \sqrt{10}}$

12. $\dfrac{X}{6+i}$ may be simplified as:

   A. $\dfrac{6X - iX}{37}$

   B. $\dfrac{6X + iX}{35}$

   C. $\dfrac{5X}{35}$

   D. $\dfrac{6X - iX}{35}$

13. Simplify $|3^2 - 8^2|$.

   A. -55
   B. -73
   C. 55
   D. 73

14. Which is not true of a geometric line?

   A. It is two dimensional.

   B. It is made up of points.

   C. It is infinite in length.

   D. It can be the intersection of two planes.

15. A rectangle measures 2A by B. If the 2A side was twice as long and the B side half as long, what would be the area of the new rectangle?

   A. 4AB square units

   B. 2AB square units

   C. 2A + 2B square units

   D. $4A + \dfrac{B}{2}$ square units

Circle your answer.

For #5–7: Justin rode his bicycle downhill to the park at 9 mph. On the return trip, he was only able to go 6 mph, and it took him 6 hours to get home.

1. Which expression is incorrect?

A. $D = \dfrac{R}{T}$

B. $T = \dfrac{D}{R}$

C. $D = RT$

D. $R = \dfrac{D}{T}$

2. Tom traveled at 60 mph for four hours. How far did he travel?

   A. 15 miles
   B. 240 miles
   C. 120 miles
   D. 1,264 miles

3. How long will it take a turtle to travel 270 feet at three feet per minute?

   A. 810 minutes
   B. 273 minutes
   C. 90 minutes
   D. 70 minutes

4. A plane travels 400 miles in 1 3/4 hours. Its average rate of travel is closest to:

   A. 175 mph
   B. 700 mph
   C. 200 mph
   D. 229 mph

5. Which value is the same for both parts of the trip?

   A. time
   B. distance
   C. rate
   D. none of the above

6. How long did the trip to the park take Justin?

   A. 4 hours
   B. 36 hours
   C. 6 hours
   D. 2 hours

7. What is the distance from Justin's home to the park?

   A. 54 miles
   B. 9 miles
   C. 18 miles
   D. 36 miles

For #8–10: Sally left point A at 6:00 a.m. and reached point B at 10:00 a.m. Jane left one hour later and arrived at point B at the same time as Sally. Jane traveled two miles an hour faster than Sally.

8. Which equation can be used to find the missing information?

    A. $(R_S)(4) = (R_S - 2)(3)$
    B. $(R_S)(4) = (R_S)(3)$
    C. $(R_S)(4) = (R_S + 2)(3)$
    D. $(R_J)(4) = (R_J + 2)(3)$

9. What was Sally's rate of travel?

    A. 4 mph
    B. 2 mph
    C. 6 mph
    D. 12 mph

10. How far did Sally and Jane travel?

    A. 24 miles
    B. 32 miles
    C. 16 miles
    D. 8 miles

11. What is the fourth term in $(2X + Y)^5$?

    A. $40X^3Y^2$
    B. $40X^2Y^3$
    C. $20X^2Y^3$
    D. $20X^4Y^1$

12. The measure of angle 1 is 132°. If A and B are parallel, what is the measure of angle 7?

    A. 42°
    B. 48°
    C. 132°
    D. unknown

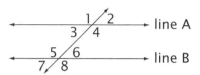

13. An angle greater than 90° and less than 180° is said to be:

    A. acute
    B. reflex
    C. right
    D. obtuse

14. What is the volume of the cylinder shown?

    A. $\pi X^2 H$
    B. $2\pi X^2 H$
    C. $\pi 4X^2 H$
    D. $\pi 16X^2 H$

15. What is the surface area of the cylinder shown above?

    A. $2\pi X^2 + 4\pi XH$
    B. $4\pi X^2 + 4XH$
    C. $8\pi X^2 + 4\pi XH$
    D. $8\pi X^2 H$

Use the figures for #1, 4, and 7.

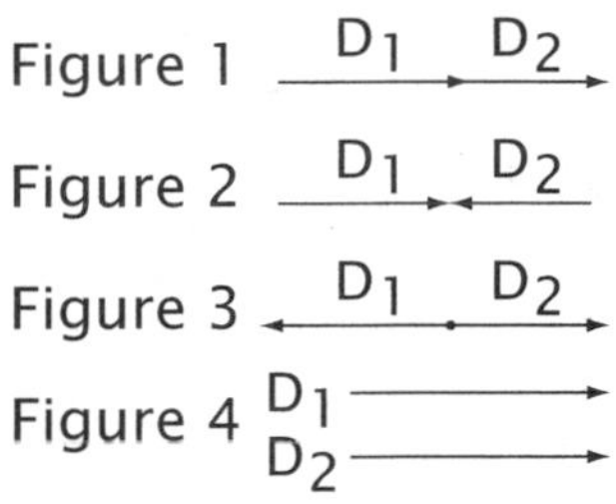

Figure 1 $\underrightarrow{D_1 \quad D_2}$

Figure 2 $D_1 \quad D_2$

Figure 3 $D_1 \quad D_2$

Figure 4 D_1
          D_2

For #1–3: Abe traveled west at 4 mph and Jack traveled east at 8 mph. They each left the starting point at the same time and traveled until they were 24 miles apart.

1. Which drawing best shows Abe's and Jack's travels?

    A. figure 1
    B. figure 2
    C. figure 3
    D. figure 4

2. Which value is the same for both Abe and Jack?

    A. time
    B. distance
    C. rate
    D. none of the above

3. How long until Abe and Jack were 24 miles apart?

    A. 2 hours
    B. 4 hours
    C. 6 hours
    D. 12 hours

For #4–6: A canoe traveled 39 miles down the river, first at a speed of 5 mph in the rapids, and then at 3 mph in slower current. It spent twice as much time in the rapids as it did in the slower current.

4. Which drawing best shows the canoe trip?

    A. figure 1
    B. figure 2
    C. figure 3
    D. figure 4

5. How long did the canoe spend in the slower current?

    A. 13 hours
    B. 3 hours
    C. 5 hours
    D. 6 hours

6. How long did the canoe spend in the rapids?

    A. 12 hours
    B. 3 hours
    C. 9 hours
    D. 6 hours

For #7–10: Victoria and Albert were 130 miles apart. Albert left first for their meeting, and Victoria left one hour later. They traveled at the same rate, and the sum of their travel times was five hours.

7. Which drawing best shows Albert and Victoria's travels?

    A. figure 1
    B. figure 2
    C. figure 3
    D. figure 4

8. At what rate of speed did Albert and Victoria travel?

    A. 15.6 mph
    B. 10.4 mph
    C. 13 mph
    D. 26 mph

9. How far did Albert travel?

    A. 26 miles
    B. 52 miles
    C. 78 miles
    D. 16 miles

10. How far did Victoria travel?

    A. 26 miles
    B. 52 miles
    C. 78 miles
    D. 32 miles

11. What is the greatest common factor of 4, 16, and 18?

    A. 2
    B. 18
    C. 4
    D. 144

12. Which of the following is not equal to $(3^3)(3^2)$?

    A. 243
    B. 3^6
    C. 3^5
    D. (3)(3)(3)(3)(3)

13. $2X^2 + 11X + 12$ divided by $2X + 3$ is equal to:

    A. $X - 4$
    B. $X + 4$
    C. $2X + 4$
    D. $X + 3$

14. Base 10:

    A. is used primarily in algebra
    B. is used only
       for metric measurement
    C. is used in the decimal system
    D. is used mostly
       for computer programs

15. Timothy walked south for six miles, and then made a right angle turn and walked west for seven miles. Approximately how many miles must he walk in order to return to his starting point by the shortest route?

    A. 85 miles
    B. 9 miles
    C. 4 miles
    D. 13 miles

Solve for the unknown by completing the square.

1. $X^2 + 6X - 6 = 0$

2. $X^2 + 4X + 1 = 0$

Solve for the unknown by using the quadratic formula.

3. $X^2 + 8X - 5 = 0$

4. $2X^2 + 3X + 6 = 0$

Find the discriminant and describe the roots of each equation. Choose from the following terms: real, imaginary, rational, irrational, equal, and unequal.

5. $X^2 + 3X = 10$

6. $X^2 + 12X + 36 = 0$

7. $4X^2 - 8X = -20$

8. $X^2 - 5X + 3 = 0$

Solve as directed. None of the unknowns equal zero.

9. Solve for W: WXY = Z

10. Solve for A: $\dfrac{TR}{A} = \dfrac{X}{B}$

11. Solve for X: $\dfrac{X}{Y} - A = 0$

12. Solve for Y: $\dfrac{1}{Y} = \dfrac{1}{T}$

Answer the questions.

13. A set of living room furniture is being discounted 20%. If the original price was $2,345, what is the new price?

14. If the tax rate is 6%, what is the total amount that must be paid for the furniture in #13?

15. The atomic weight of carbon (C) is 12 and that of oxygen (O) is 16. What is the percentage of carbon in CO_2?

16. The ratio of blue to green is six to eight. If 36 are blue, how many are green?

17. Jim threw the basketball 56 times. His ratio of baskets to misses was five to nine. How many baskets did he make?

18. Express 400 ounces as pounds.

19. There are .62 miles in one kilometer. How many miles are there in eight kilometers?

20. Greg traveled at 50 mph for six hours. How far did he travel?

21. How long will it take Cody to travel 280 miles if he travels at 20 mph?

22. Traveling at a rate of 60 mph, Angie took one hour to get to the city this morning. Coming home by the same route, she was able to travel at only one-third of her morning rate because of traffic. How far is it to the city? How long did it take Angie to drive home?

23. Lewis left camp at 5:00 a.m. and reached the lake at 10:00 a.m. Clark left camp one hour later and reached the lake at the same time. Clark paddled his canoe two miles an hour faster than Lewis. How fast did each paddle?

24. Joseph and Devan were 230 miles apart. Joseph started toward their meeting place at 1:00 p.m. traveling at 35 mph. Devan left at 3:00 p.m. and traveled 10 mph faster than Joseph. When did they meet?

25. How far did each person in #24 travel?

Circle your answer.

1. Which equation is in the slope-intercept form?

    A. Y = 13X – 2
    B. 4Y = 2X + 1
    C. 3X + 4Y = 2
    D. 2X + Y = 0

2. A line with a negative slope:

    A. has only negative coordinates
    B. slopes up to the right
    C. slopes down to the right
    D. none of the above

For #3–5, match the given equation with the most probable line.

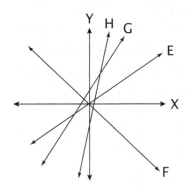

3. Y = 5X – 3

    A. E
    B. F
    C. G
    D. H

4. Y = –2X

    A. E
    B. F
    C. G
    D. H

5. Y = 3X + 2

    A. E
    B. F
    C. G
    D. H

6. Given a slope of 3 through the point (1, 2), find the Y-intercept of the line.

    A. 1
    B. 5
    C. –1
    D. –5

7. The formula for calculating slope is:

    A. $\dfrac{Y_2 - Y_1}{X_2 - X_1}$

    B. $\dfrac{Y_2 + Y_1}{X_2 + X_1}$

    C. $\dfrac{X_2 + X_1}{Y_2 + Y_1}$

    D. $\dfrac{X_1 - Y_1}{X_2 - Y_2}$

8. Given a line through points (–1, 5) and (1, 1), what is the equation in slope-intercept form?

    A. Y + 2X – 3 = 0
    B. Y = –3X + 2
    C. Y = 3X – 2
    D. Y = –2X + 3

9. What is the answer to #8 written as the standard equation of a line?

    A. Y = 3X + 2
    B. 2X + Y = 3
    C. Y – 2X + 3 = 0
    D. Y – 3X + 2 = 0

10. Given a line through points (–1, 1) and (3, –2), which equation is true?

    A. Y = –3/4 X + 1/4
    B. Y = –3/2 X – 1/2
    C. Y = –1/4 X + 3/4
    D. Y = 3/4 X – 1/4

11. Solve for X: $AX^2 + BX + C = 0$.

    A. $X = \dfrac{-B \pm \sqrt{B^2 - (4)(A)(C)}}{2(A)}$

    B. $X = \dfrac{B \pm \sqrt{B^2 - (4)(A)(C)}}{2(A)}$

    C. $X = \dfrac{-B \pm \sqrt{B^2 + (4)(A)(C)}}{2(A)}$

    D. $X = \dfrac{-B + \sqrt{B^2 - (4)(A)(C)}}{2}$

12. If the discriminant is negative, the roots of the equation are:

    A. real, rational, unequal
    B. real, rational, equal
    C. real, irrational, unequal
    D. imaginary

13. The final bill at the restaurant was $68.15, including a 15% tip and a 5% tax. What was the cost of the food?

    A. $43.97
    B. $56.79
    C. $13.63
    D. $66.15

14. The distance from earth to the sun is 93 million miles. If a space ship travels 30 thousand miles a day, how many days will it take to reach the sun?

    A. 3.1×10^3
    B. 3.1×10^{10}
    C. 3.1×10^{-3}
    D. 2.8×10^{11}

15. About how many years will it take the spaceship described in #14 to reach the sun?

    A. 82.2 years
    B. 10 years
    C. 8.2 years
    D. 8.5 years

Circle your answer.

1. Two lines that have the same slope and different intercepts are:

    A. perpendicular
    B. parallel
    C. reciprocals
    D. none of the above

2. The slopes of two perpendicular lines are:

    A. reciprocals
    B. the same
    C. the same except for the sign
    D. negative reciprocals

3. Which line is parallel to $Y = 2X + 5$?

    A. $2Y = 4X + 3$
    B. $3Y + 6X = 10$
    C. $Y = -2X + 4$
    D. none of the above

4. Which line is perpendicular to $Y = 1/4 X + 2$?

    A. $Y = -1/4 X + 2$
    B. $Y = 4X + 3$
    C. $Y = -2X + 2$
    D. none of the above

5. Which one is an equation of a line parallel to $Y = 2X + 3$, passing through $(2, -2)$?

    A. $Y = -2X - 6$
    B. $Y = -1/2 X - 1$
    C. $Y = 2X - 6$
    D. $Y = 2X + 6$

6. Which one is an equation of a line perpendicular to $Y = 2X + 3$, passing through $(2, -2)$?

    A. $Y = 2X - 6$
    B. $Y = -1/2 X - 1$
    C. $Y = -1/2 X - 6$
    D. $Y = 2X + 6$

7. When multiplying both sides of an inequality by -1:

    A. the inequality sign changes direction
    B. the inequality sign stays the same
    C. the solution of the equation changes
    D. the less than or equals sign becomes an equals sign

For #8–10:

Figure 1   Figure 2   Figure 3   Figure 4

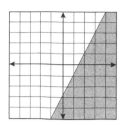 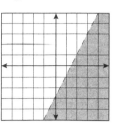 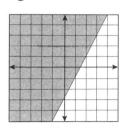 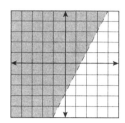

8. Which graph shows Y > 2X – 3?

    A. figure 1
    B. figure 2
    C. figure 3
    D. figure 4

9. Which graph shows Y ≥ 2X – 3?

    A. figure 1
    B. figure 2
    C. figure 3
    D. figure 4

10. Which graph shows Y < 2X – 3?

    A. figure 1
    B. figure 2
    C. figure 3
    D. figure 4

11. Which of these figures does not have two pairs of parallel sides?

    A. parallelogram
    B. square
    C. trapezoid
    D. rhombus

12. The area of an enclosure is $X^4 - 1$. If it is divided into sections, each with an area of $X^2 - 1$, how many sections are made?

    A. $X^2 - 2$
    B. $X + 1$
    C. $X^2 + 1$
    D. X^2

13. If two parallel lines are cut by a transversal, which is not true:

    A. Corresponding angles are congruent.
    B. Alternate interior angles are congruent.
    C. Alternate exterior angles are congruent.
    D. Supplementary angles are congruent.

14. Simplify: $3|-4 - 2| + 1 - 2(3 - 6)^2$

    A. -35
    B. 1
    C. 51
    D. -11

15. $8^{-1/3}$ is the same as:

    A. $1/2$
    B. $\sqrt[3]{-8}$
    C. 2
    D. $-8/3$

Circle your answer.   For #5–10:

1. The distance formula for two
   points on a graph is based on:

   A. the Pythagorean theorem
   B. quadratic formula
   C. midpoint formula
   D. none of the above

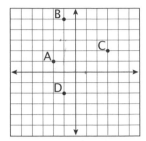

2. The distance formula may be
   written as:

   A. $D = (X_B - X_A)^2 + (Y_B - Y_A)^2$

   B. $D = \sqrt{(X_B - X_A) + (Y_B - Y_A)}$

   C. $D = \sqrt{(X_B - X_A)^2 + (Y_B - Y_A)^2}$

   D. $D = (X_B + X_A)^2 - (Y_B + Y_A)^2$

3. In order to find the coordinates
   of the midpoint, add the
   X-coordinates of the endpoints,
   and then the Y-coordinates of the
   endpoints and:

   A. multiply each sum by two
   B. divide each sum by two
   C. add the sums
   D. subtract the sums

4. Which statement is never true of
   the distance between two points on
   a graph?

   A. It is a negative number.
   B. It is a rational number.
   C. It is a whole number.
   D. It is an irrational number.

5. Find the distance between A and C.

   A. 26
   B. $2\sqrt{6}$
   C. $\sqrt{2}$
   D. $\sqrt{26}$

6. Find the distance between B and C.

   A. $\sqrt{13}$
   B. 5
   C. $\sqrt{53}$
   D. 25

7. Find the distance between A and D.

   A. $\sqrt{10}$
   B. 10
   C. 8
   D. $2\sqrt{2}$

8. Find the midpoint between B
   and D.

   A. (–2, 3)
   B. (3/2, –1)
   C. (–1, 3/2)
   D. (0, 7/2)

9. Find the midpoint between D and C.

    A. (2, 2)
    B. (1, 0)
    C. (2, 1)
    D. (2, 4)

10. Find the midpoint between A and B.

    A. (-3, 6)
    B. (-1/2, -2)
    C. (-3/2, 3)
    D. (3, -3/2)

11. Solve for C: $\dfrac{AB}{C} = \dfrac{Y}{X}$

    A. $C = \dfrac{ABY}{X}$

    B. $C = \dfrac{ABX}{Y}$

    C. $C = \dfrac{XY}{AB}$

    D. $C = \dfrac{AB}{XY}$

12. The customers preferred apples to bananas by a ratio of four to three. A total of 51 customers preferred bananas. How many customers were there altogether?

    A. 68
    B. 191
    C. 89
    D. 119

13. In order to find the area of a trapezoid, first:

    A. find the average length of the bases

    B. take base times height and divide by two

    C. use the length of the longest side as the height

    D. square the height

14. A pole is 192.5 cm tall. If .4 in equals one cm, how many feet tall is the pole?

    A. 77 ft
    B. 48 ft
    C. 6.4 ft
    D. 25.7 ft

15. A right triangle with sides of 8 and 10 would have a hypotenuse of:

    A. $3\sqrt{2}$

    B. 15

    C. 164

    D. $2\sqrt{41}$

Circle your answer.

1. Which is the equation of a circle?

   A. $4X^2 + 9Y^2 = 36$
   B. $Y = 2X^2$
   C. $XY = 6$
   D. $X^2 + Y^2 = 36$

2. Which is the equation of an ellipse?

   A. $4X^2 + 9Y^2 = 36$
   B. $Y = 2X^2$
   C. $XY = 6$
   D. $X^2 + Y^2 = 36$

3. If $X^2 + Y^2 = Z^2$, then Z represents:

   A. the diameter of a circle
   B. one coordinate
      of the midpoint
   C. the radius of a circle
   D. none of the above

4. $X^2 - 6X + Y^2 + 10Y = -18$, rewritten
   as the equation of a circle, is:

   A. $(X - 3)^2 + (Y + 5)^2 = 9^2$
   B. $X^2 - 6X + 18 = -Y^2 - 10Y$
   C. $(X - 3)^2 + (Y + 5)^2 = 4^2$
   D. $(X^2 - 3) + (Y^2 + 5) = 16$

5. What are the coordinates of the
   center of $(X - 2)^2 + (Y + 3)^2 = 49$?

   A. $(2, -3)$
   B. $(-2, 3)$
   C. $(0, 0)$
   D. $(-5, 7)$

6. The formula for an ellipse can be
   distinguished from that of a circle
   by its:

   A. irrational radius
   B. unequal coefficients
   C. negative coefficients
   D. equal coefficients

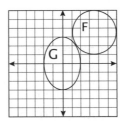 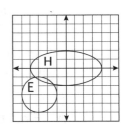

7. Which figure represents
   $(X + 3)^2 + (Y + 3)^2 = 4$?

   A. E
   B. F
   C. G
   D. H

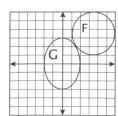 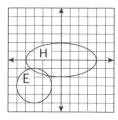

8. Which figure represents
$9X^2 + 4Y^2 = 36$?

    A. E
    B. F
    C. G
    D. H

9. Given $2(X - 1)^2 + 3(Y + 1)^2 = 52$,
the coordinates of the center are:

    A. (–1, 1)
    B. (3, –2)
    C. (–2, 3)
    D. (1, –1)

10. To find the extremities of
an ellipse:

    A. Find the square root
       of each term.
    B. Use the X and Y extremities
       for one point.
    C. Find the factors
       of the constant.
    D. Make the X term, and then
       the Y term, equal to zero.

11. Which expression has the smallest
value?

    A. $-3^2 \div 3 + 6$
    B. $(3)^2 \div 3 + 5$
    C. $-(3)^2 \div 3 + 5$
    D. $(-3)^2 \div 3 + 4$

12. $\dfrac{2}{X} + \dfrac{6}{3X} - X^{-1} =$

    A. $\dfrac{5X + 3}{3X}$

    B. $\dfrac{3}{X}$

    C. $\dfrac{8}{3X} - X^{-1}$

    D. $\dfrac{12 + X}{3X}$

13. The intersection of two lines is a(n):

    A. plane
    B. point
    C. line segment
    D. union

14. David and Daniel each walked
the same distance to work. David
walked 4 mph and Daniel walked 6
mph. Daniel's walk took one hour
less than David's. How far is it to
work?

    A. 12 miles
    B. 2 miles
    C. 18 miles
    D. 8 miles

15. Each side of a rectangular figure is
tripled in length. What happens to
the value of the area?

    A. It is doubled.
    B. It is tripled.
    C. It is nine times as great.
    D. Results will vary.

Circle your answer.   For 6–10:

1. Which of the following is not the
    equation of a parabola?

    A. $XY = 2$
    B. $X = Y^2$
    C. $Y = -X^2$
    D. $Y = 1/2X^2$

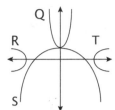
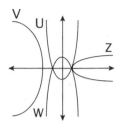

2. In which direction does the graph
    of $Y = 2X^2$ open?

    A. up
    B. down
    C. right
    D. left

6. Which graph represents
    $Y = 2X^2 + 1$?

    A. Q
    B. U
    C. Y
    D. Z

3. In which direction does the graph
    of $X = -3Y^2$ open?

    A. up
    B. down
    C. right
    D. left

7. Which graph represents
    $X = -1/3\ Y^2 - 2$?

    A. S
    B. T
    C. V
    D. W

4. Which of the following equations
    yields the narrowest parabola?

    A. $Y = -1/4\ X^2$
    B. $X = -3Y^2$
    C. $Y = 2X^2$
    D. $X = Y^2$

8. Which graph represents
    $Y = -1/3\ X^2 + 1$?

    A. S
    B. T
    C. V
    D. W

5. What are the coordinates of the
    vertex of $Y = X^2 + 2$?

    A. (2, 0)
    B. (-2, 0)
    C. (0, 2)
    D. (0, -2)

9. Which graph represents $X = Y^2 + 1$?

    A. U
    B. T
    C. Q
    D. Z

For #13–15:

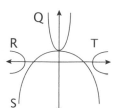 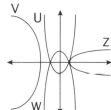

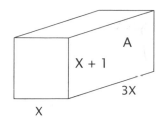

10. Which graph represents $Y = X^2 - 1$?

    A. S
    B. U
    C. R
    D. W

13. What is the perimeter of side A?

    A. $4X + 1$
    B. $3X^2 + 1$
    C. $8X + 2$
    D. $3X^2 + 3X$

11. The vertex of $\angle RST$ is:

    A. R
    B. S
    C. T
    D. not enough information

14. What is the surface area of the rectangular solid?

    A. $7X^2 + 4X$ units^2
    B. $3X^2 + 3X$ units^2
    C. $14X^2 + 8X$ units^2
    D. $3X^3 + 3X^2$ units^2

12. Oscar finished the 44 km trip in 6 hours. He traveled part of the distance at 4 km/h and part at 9 km/h. How long did he travel at 9 km/h?

    A. 2 hours
    B. 4.9 hours
    C. 5 hours
    D. 4 hours

15. What is the volume of the rectangular solid?

    A. $3X^3 + 3X^2$ units^3
    B. $3X^2 + 3X$ units^3
    C. $14X^2 + 8X$ units^3
    D. $4X^2 + 4X$ units^3

Circle your answer.

1. In the equation of a parabola, the middle term:

   A. moves the vertex on the X-axis
   B. moves the vertex on the Y-axis
   C. determines the direction in which the parabola opens
   D. moves the parabola off the X- or Y-axis

2. The lowest point of a positive parabola is the:

   A. maxima
   B. axis of symmetry
   C. endpoint
   D. minima

3. The highest point of a negative parabola is the:

   A. maxima
   B. axis of symmetry
   C. endpoint
   D. minima

4. The axis of symmetry of a parabola:

   A. determines the lowest point of the parabola
   B. determines the highest point of the parabola
   C. divides the parabola into symmetrical halves
   D. always lies on the X- or Y-axis

5. The formula used to find the X-coordinate of the axis of symmetry of a vertical parabola is:

   A. $X = \dfrac{B}{2A}$

   B. $X = \dfrac{-B}{2A}$

   C. $X = \dfrac{2B}{A}$

   D. $X = \dfrac{-B}{2}$

6. What are the coordinates of the vertex of $Y = X^2 - 8X + 1$?

   A. (−4, 49)
   B. (4, −15)
   C. (−15, 4)
   D. (−8, 1)

7. What are the coordinates of the vertex of $Y = -3X^2 + 6X$?

   A. (−1, −9)
   B. (3, 1)
   C. (1, 3)
   D. (−1, −3)

8. What are the coordinates of the vertex of $Y = 1/2\ X^2 - 4X$?

   A. (4, −8)
   B. (−4, 24)
   C. (4, −12)
   D. (2, −2)

9. What is the largest rectangular area that can be enclosed with 240 feet of fence?

   A. 14,400 ft^2
   B. 900 ft^2
   C. 7,200 ft^2
   D. 3,600 ft^2

10. A 96" length of trim is available to go around three sides of a window opening. What are the dimensions of the window with the largest possible opening?

    A. 24" x 24"
    B. 24" x 48"
    C. 48" x 48"
    D. 1,152"

11. $3\sqrt{-27} - 4\sqrt{-8} =$

    A. $3i\sqrt{3} - 4i\sqrt{2}$

    B. $9i\sqrt{3} - 8i\sqrt{2}$

    C. $-i\sqrt{-19}$

    D. $72i\sqrt{6}$

12. $9i\sqrt{-64}$

    A. -72
    B. 72
    C. 72i
    D. -72i

13. Put $\dfrac{2i}{8+5i}$ in standard form.

    A. $\dfrac{16i+10}{89}$

    B. $\dfrac{16i-10}{89}$

    C. $\dfrac{16i+10}{39}$

    D. $16i - 10i^2$

14. Solve for X: $2X^2 + 2X - 5 = 0$

    A. $X = \dfrac{-1\pm 3i}{2}$

    B. $X = \dfrac{-2\pm\sqrt{11}}{4}$

    C. $X = \dfrac{1\pm 3i}{2}$

    D. $X = \dfrac{-1\pm\sqrt{11}}{2}$

15. When the discriminant is negative, the roots of a quadratic equation are imaginary because:

    A. The roots of a quadratic equation cannot be negative.
    B. The square root of a negative number is imaginary.
    C. Negative numbers cannot be used in the denominator of a quadratic equation.
    D. None of the above; a negative discriminant yields negative roots.

92%

Circle your answer.

1. Which of the following is not the equation of a hyperbola?

   A. $XY = -6$

   B. $X = \dfrac{16}{Y}$

   C. $9X^2 - 4Y^2 = 36$

   D. $2Y^2 + X^2 = 16$

2. Which is true of an inverse relationship?

   A. One variable increases while the other decreases.
   B. One variable increases while the other remains the same.
   C. One variable remains the same and the other decreases.
   D. Both variables increase at the same time.

3. On the graph of an inverse relationship:

   A. All the points fall in the same quadrant.
   B. The X-intercept has a very large value.
   C. The X-intercept has a very small value.
   D. The lines approach, but do not touch the axis.

4. In the equation $XY = 6$, if the value of $X = -1$, the value of Y is:

   A. 6
   B. ± 6
   C. -6
   D. -1/6

5. In the equation $X^2 - Y^2 = 8$, if the value of $Y = -1$, the value of X is:

   A. -3
   B. +3
   C. ± 3
   D. 9

6. The graph for $XY = -1$ lies in which quadrants?

   A. I and III
   B. I and II
   C. III and IV
   D. II and IV

For #7–10.

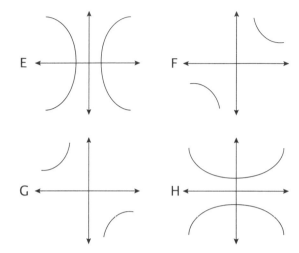

7. Which equation corresponds to graph E?

    A. $9X^2 + 4Y^2 = 36$
    B. $9X^2 - 4Y^2 = 36$
    C. $XY = 36$
    D. $Y = 9X^2 + 36$

8. Which equation corresponds to graph F?

    A. $XY = 6$
    B. $X^2 - Y^2 = 6$
    C. $XY = -6$
    D. $X = Y^2 + 6$

9. Which equation corresponds to graph G?

    A. $XY = 6$
    B. $X^2 - Y^2 = 6$
    C. $XY = -6$
    D. $X = Y^2 + 6$

10. Which equation corresponds to graph H?

    A. $XY = 18$
    B. $XY = -18$
    C. $2Y^2 + X^2 = 18$
    D. $2Y^2 - X^2 = 18$

11. Which line is perpendicular to $Y = 1/2\ X + 4$ through the point (0, 0)?

    A. $Y = 2X$
    B. $Y = -1/2\ X$
    C. $Y = -2X$
    D. $X = -2Y$

12. What is the midpoint between (–5, –3) and (7, –6)?

    A. (1, –9/2)
    B. (–9/2, 1)
    C. (–6, 3/2)
    D. (6, 3/2)

13. Which of the following can not be formed by a conic section?

    A. ellipse
    B. square
    C. hyperbola
    D. parabola

14. The wholesale price of a chair is $254.00 and the retail price is $299.00. What percent of the retail price is the markup?

    A. 17.7%
    B. 85%
    C. 6.6%
    D. none of the above

15. The atomic weight of sodium (Na) is 23 and the atomic weight of chlorine (Cl) is 35. What is the approximate percentage of sodium in NaCl?

    A. 66%
    B. 60%
    C. 15%
    D. 40%

Circle your answer.

1. Which methods may be used to solve for two equations at the same time?

   I. substitution, II. translation, III. elimination, IV. graphing

   A. I and II only
   B. I, II and III only
   C. I, III, and IV only
   D. none of the above

2. Which of the following is a possible first step in solving $X^2 + Y^2 = 16$ and $Y = 2X - 1$?

   A. Substitute $2X - 1$ for Y in the first equation.
   B. Substitute $2X - 1$ for X in the first equation.
   C. Divide $X^2 + Y^2 = 16$ by two.
   D. Multiply $Y = 2X - 1$ by 16.

3. Find the solution of $XY = 6$ and $3X + 2Y = 12$.

   A. (3, 2)
   B. (–2, –3)
   C. (–3, 2)
   D. (2, 3)

4. Given $Y = X - 3$ and $X^2 + Y^2 = 9$, how many solutions are there?

   A. one
   B. two
   C. three
   D. four

5. What is one solution for the two equations given in #4?

   A. (–3, 0)
   B. (3/2, 9/2)
   C. (0, –3)
   D. (–3/2, 3/2)

6. The largest number of possible solutions for the equations of a circle and a hyperbola is:

   A. one
   B. two
   C. three
   D. four

7. Find the solution(s) of $Y = X^2 + 2$ and $X^2 + Y^2 = 4$.

   A. (0, 2) and $(-3, \sqrt{5})$
   B. (0, 2)
   C. (2, 0)
   D. (3, 1) and (2, 0)

8. Find the solution(s) of $X^2 - Y^2 = 16$ and $X^2 + Y^2 = 34$.

   A. (5, ± 3) and (–5, ± 3)
   B. (–5, 3)
   C. (5, 3) and (3, 5)
   D. (5, 3) and (–5, –3)

9. Why is it a good idea to check all roots in the given equations?

    A. Solutions containing radicals are not allowed.
    B. Fractional solutions cannot be graphed.
    C. Some roots may require imaginary numbers to solve the equations.
    D. Very large or small roots may not work.

10. Graphing two equations:

    A. is not useful for graphs based on conic sections
    B. gives the most exact solutions of the equations
    C. gives a good estimate of the answer
    D. may show impossible solutions

11. What is the area of the largest rectangular space you can enclose with 176 ft of fence?

    A. 1,936 ft^2
    B. 88 ft^2
    C. 44 ft^2
    D. 7,744 ft^2

12. If you divide A by BC, it is the same as D divided by E. What is the value of B?

    A. $\dfrac{AE}{CD}$

    B. $\dfrac{A}{C} - \dfrac{D}{E}$

    C. ACDE

    D. EC

13. 1,200,000,000 bacteria each had 3,000,000 offspring. How many new bacteria are there?

    A. 3.6×10^{54}
    B. 4.2×10^{15}
    C. 3.6×10^{15}
    D. 1.203×10^{9}

14. Jeff traveled 18 miles at an average speed of 15 mph. Coming home, he traveled 3 mph slower. How long did the return trip take?

    A. 1.2 hours
    B. 1.5 hours
    C. 1 hour
    D. 3.6 hours

15. A rectangular park measures 3 mi by 2 mi. What is the area of the park in kilometers if 1 mi equals 1.6 km?

    A. 3.75 km^2
    B. 9.6 km^2
    C. 8 km^2
    D. 15.36 km^2

1. Klara has 11 coins, all either dimes or nickels. If she has a total of $.80, how many dimes does she have?

   A. 6
   B. 5
   C. 7
   D. 3

2. An ice cream cone cost $1.15. It was paid for with seven coins consisting of quarters and dimes. How many quarters were used?

   A. 3
   B. 4
   C. 7
   D. 5

3. Taylor has 25 coins consisting of pennies and nickels. The sum of his coins is $.57. How many of each coin does he have?

   A. 8 pennies, 17 nickels
   B. 16 pennies, 9 nickels
   C. 17 pennies, 8 nickels
   D. 7 pennies, 5 nickels

4. Find three consecutive even integers such that three times the first, plus the second, plus two, equals three times the third.

   A. 8, 9, 10
   B. 10, 11, 12
   C. 8, 10, 12
   D. 4, 6, 8

5. Which of the following represents three consecutive odd integers?

   A. N, N + 3, N + 5
   B. N, N + 1, N + 2
   C. N, N – 2, N – 4
   D. N, N + 2, N + 4

6. Find three consecutive integers such that four times the first, plus two times the second, is equal to four times the third.

   A. 3, 4, 5
   B. 3, 5, 7
   C. 3, 2, 1
   D. 4, 5, 6

7. Find three consecutive odd integers such that 10 times the first, plus 10 times the second, equals 10 more than 10 times the third. What is the third number?

   A. 4
   B. 5
   C. 6
   D. 7

8. A chemist needs 90 ml of a 10% acid solution. He has on hand a 20% acid solution and an 8% acid solution. How much of the 20% solution should he use?

   A. 75 ml
   B. 15 ml
   C. 28 ml
   D. 14 ml

9. A farmer wants 150 pounds of fertilizer that is 12% nitrogen. He has already purchased fertilizer that is 50% nitrogen and some that is 5% nitrogen. How much of the 5% mixture should he use?

   A. 360 lb
   B. 126.7 lb
   C. 23.3 lb
   D. 55 lb

10. How many pounds of the 50% mixture should the farmer in #9 use?

   A. 360 lb
   B. 126.7 lb
   C. 23.3 lb
   D. 95 lb

11. Leaving the Forum at 12:00 noon, Caesar traveled west at 4 mph and Brutus traveled east at 6 mph. At what time were they 20 miles apart?

   A. 4:00 p.m.
   B. 5:00 p.m.
   C. 3:20 p.m.
   D. 2:00 p.m.

12. Which of the following is the equation of a parabola?

   A. $Y = X^2 + 3$
   B. $X + Y = 3^2$
   C. $XY = 3$
   D. $X^2 + Y^2 = 3$

13. What is the sum of the interior angles of the regular polygon shown?

   A. 1,440°
   B. 2,160°
   C. 1,080°
   D. 135°

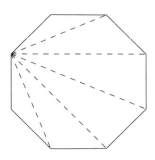

14. I have 720 ft of fencing to enclose a rectangular area. One side will be protected by an existing fence. What are the dimensions of the largest area that can be enclosed?

    A. 64,800 ft^2
    B. 180' by 180'
    C. 180' by 360'
    D. 32,400 ft^2

15. Simplify: $\dfrac{\dfrac{3}{X}}{\dfrac{2}{X+1}}$

    A. $\dfrac{6}{X^2+X}$

    B. $\dfrac{3X+3}{2X}$

    C. $\dfrac{6}{2X}$

    D. $\dfrac{3X+1}{2X}$

Circle your answer.

1. In two years, Roger will be 3 1/2 times the age of Shane. Three years ago, Roger was six times Shane's age. How old is Shane now?

   A. 8
   B. 10
   C. 35
   D. 33

2. Eight years ago, Pat was one-third the age of Karen. In two years, Karen will be twice the age of Pat. How old is Karen?

   A. 18
   B. 38
   C. 30
   D. 40

3. Five years ago, Cameron was one-half the age of Drew. In three years, Cameron will be five-sixths of Drew's age. How old is Drew?

   A. 10
   B. 7
   C. 9
   D. 12

4. In 30 years, Sarah will be three-fifths the age of her dad. Five years ago, her dad was four times Sarah's age. How old is Sarah now?

   A. 45
   B. 15
   C. 20
   D. 10

5. In 10 years, Rick will be four times the age of Chance. Today Rick is 10 times the age of Chance. How old was Chance two years ago?

   A. 5
   B. 13
   C. 48
   D. 3

6. Which equation is not true for boat-in-the-current problems?

   A. $R_{downstream} = R_{boat} + R_{water}$

   B. $R_{upstream} = R_{boat} - R_{water}$

   C. $R_{upstream} = R_{water} - R_{boat}$

   D. $D_{downstream} = R_{downstream} \times T_{downstream}$

7. A boat can go 42 miles downstream in the same time it takes to go 18 miles upstream. The rate of the water is 4 mph. What is the rate of the boat?

   A. 18 mph
   B. 3 mph
   C. 6 mph
   D. 10 mph

8. Captain Smith paddled 12 miles upstream in the same time it took him to paddle 60 miles downstream. The rate of the current was 2 mph. What was his total travel time?

   A. 12 hours
   B. 14 hours
   C. 28 hours
   D. 24 hours

9. A boat has a speed of 10 mph without any current. It went downstream for two hours, then returned to the starting point in five hours. What is the speed of the current?

   A. 4.3 mph
   B. 3.75 mph
   C. 5.7 mph
   D. 10 mph

10. How far did the boat in #9 travel downstream?

    A. 5 miles
    B. 57.2 miles
    C. 28.6 miles
    D. 20 miles

11. Vontoria noted that two triangles were similar. Which is not true of the two triangles?

    A. Corresponding angles are equal.
    B. Corresponding sides are equal.
    C. Corresponding sides all form the same ratio.
    D. The sides of one triangle may all be multiplied by a constant factor to determine the sides of the second.

12. Which is an equation of a line with a negative slope?

    A. $-Y = -X + 4$
    B. $Y = -X^2 - 4$
    C. $XY = -4$
    D. $-Y = X + 4$

13. What number should go in the
    blank to complete the square?

    $X^2 + 3X +$ _____

    A. 9
    B. 9/4
    C. 9/2
    D. 3/2

14. Zarah had 14 coins consisting
    of nickels and quarters. She
    gave three nickels to her brother
    Raleigh. How many nickels does
    she have now if she had $1.90
    at first?

    A. 8
    B. 6
    C. 5
    D. 3

15. The area of a square is $9X^2Y$.
    What is the length of one side
    of the square?

    A. $3X\sqrt{Y}$
    B. 3XY
    C. $2.25\ X^2Y$
    D. none of the above

Circle your answer.

1. In order to solve for three variables, the minimum number of equations needed is:

   A. one
   B. two
   C. three
   D. four

2. To solve equations with three variables, start by:

   A. substituting trial values for the variables in one equation
   B. adding all three equations together
   C. using elimination to get two equations with the same two variables in each
   D. isolating X in one of the equations

3. Which is true of equations with three variables?

   A. None of the variables may have the same value.
   B. Substitution and elimination are the main tools for solving them.
   C. The values for all three variables may be found using only two of the equations.
   D. They may be solved by plotting lines on a graph.

4. Given: A) $2X + 3Y - Z = 11$ and B) $3X - Y + 2Z = 4$, how would you prepare to eliminate Z?

   A. Multiply A by 2.
   B. Multiply A and B by 2.
   C. Multiply B by 3.
   D. Multiply A by (–2) and B by (–1).

For #5–7:

   A $4X + 2Y + 3Z = 10$
   B $2X - 2Y + Z = -6$
   C $X + 3Y - 4Z = 22$

5. What is the value of X that satisfies all three equations?

   A. –2
   B. 3
   C. 4
   D. 2

6. What is the value of Y that satisfies all three equations?

   A. 2
   B. 3
   C. –3
   D. 4

7. What is the value of Z that satisfies all three equations?

   A. 2
   B. 4
   C. –2
   D. –4

For #8–10:

A. $2X - Y + 4Z = 5$
B. $X + 3Y - 2Z = 8$
C. $3X + 3Y - Z = 9$

8. What is the value of X that satisfies all three equations?

   A. –1
   B. 1
   C. 5
   D. –3

9. What is the value of Y that satisfies all three equations?

   A. 1
   B. 3
   C. 4
   D. 5

10. What is the value of Z that satisfies all three equations?

    A. –3
    B. 3
    C. 1
    D. 2

11. $X = \dfrac{3}{Y}$ is the equation of a(n):

    A. line
    B. ellipse
    C. parabola
    D. hyperbola

12. The maximum number of possible solutions for $Y = AX + B$ and $X^2 + Y^2 = C$ is:

    A. 1
    B. 2
    C. 3
    D. 4

13. $(\sqrt[3]{27}\,)^{-1}$ is equal to:

    A. 3
    B. 1/3
    C. –3
    D. $\sqrt{3}$

14. There are 1,368 grams of CS_2. What is the mass of the carbon if the atomic weight of C is 12, and the atomic weight of S is 32?

    A. 216 g
    B. 373 g
    C. 513 g
    D. 256.5 g

15. $\square = 2$, $A = 2 + X$, and $D = -3$. What is the value of $\square(A - D)^2$ in simplest form?

    A. $X^2 + 10X + 25$
    B. $4X^2 + 20X + 100$
    C. $2X^2 + 20X + 50$
    D. $2X^2 - 10X + 2$

Circle your answer.

1. Vectors have two components, which are:

   A. magnitude and measure
   B. direction and magnitude
   C. magnitude and force
   D. measure and weight

2. The combination of two vectors is called the:

   A. magnitude
   B. sine
   C. resultant
   D. inverse

3. The sine ratio is the same as:

   A. opposite over hypotenuse
   B. adjacent over hypotenuse
   C. opposite over adjacent
   D. hypotenuse over opposite

4. Sam went from his home to school by walking 60 yards due east, and then 80 yards due north. How far is it from his home to the school in a straight line?

   A. 10 yards
   B. 10,000 yards
   C. 100 yards
   D. 6,400 yards

5. In #4 above, what angle (from due east) would Sam have to walk in order to get to school from his house by the most direct route?

   A. 53.1°
   B. .02°
   C. 1.3°
   D. 70°

6. The tangent ratio is the same as:

   A. opposite over hypotenuse
   B. adjacent over hypotenuse
   C. opposite over adjacent
   D. hypotenuse over opposite

7. A ship traveled 50 miles east and 70 miles north. What is the distance part of the resultant vector?

   A. 11 miles
   B. 7,400 miles
   C. 50 miles
   D. 86 miles

8. In #7, what is the direction part of the resultant vector?

   A. 35.5°
   B. 54.5°
   C. 45.5°
   D. 9.0°

9. The first vector is (+2) on the X-axis and (+5) on the Y-axis. The second vector is (+3) on the X-axis and (−1) on the Y-axis. What is the final vector?

 A. +7, +2
 B. +5, +6
 C. +5, +4
 D. +6, −5

10. Give the magnitude and direction of the final vector in #9.

 A. 38, 6.4°
 B. 7.8, 50.3°
 C. 7.3, 15.9°
 D. 6.4, 38.7°

11. $4i - 3i^2 =$

 A. 7i
 B. i
 C. 4i + 3
 D. 4i − 3

12. All squares are:

 A. triangles
 B. congruent
 C. rectangles
 D. circles

13. $\dfrac{16^{-\frac{1}{2}} \cdot 3^{-2}}{2^{-2}}$

 A. 1/5
 B. 1/4
 C. 9
 D. 1/9

14. $i^{401} =$

 A. −1
 B. 0
 C. i
 D. $\sqrt{i}$

15. $\dfrac{B^{\frac{1}{3}} B^4}{B^{-2}}$

 A. $B^{13/3}$
 B. $B^{4/3}$
 C. $B^{19/3}$
 D. $B^{2/3}$

Answer the questions.

1. A line passes through the points (3, 3) and (-1, 1). Give its equation in slope-intercept form.

2. What is true of two different lines that have the same slope?

3. What is true of two different lines whose slopes are negative reciprocals of each other?

4. When both sides of an inequality are multiplied by negative one, what happens to the inequality sign?

5. Given the points (-2, 5) and (2, 2), what is the distance between them?

6. What is the midpoint between the two points given in #5?

Name the figure represented by each formula below. Sketch each figure on a graph.

7. $X^2 + Y^2 = 4$

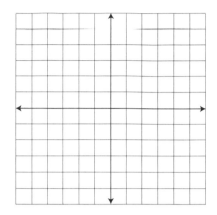

8. $4X^2 + Y^2 = 16$

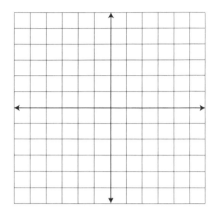

9. $Y = X^2$

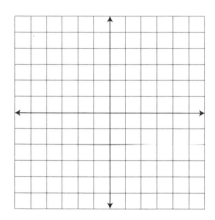

10. $XY = 6$

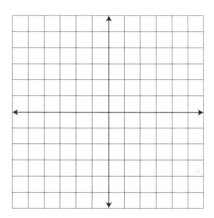

Find the vertex (maxima or minima).

11. $Y = X^2 - 4X + 2$

Find the solutions for each pair of equations. Sketch a graph of each equation, and show the solutions.

12. $X^2 + Y^2 = 9$
    $Y = X + 3$

13. $Y = X^2$
    $XY = 8$

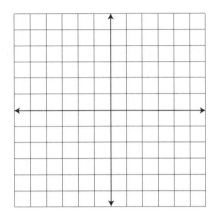
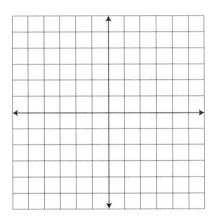

Answer the questions.

14. I have 12 coins in my pocket. They are all either dimes or nickels. The value of the coins is $.85. How many of each coin do I have?

15. Find three consecutive integers such that two times the first, plus the second, equals the third integer plus nine.

16. A scientist needs three liters of a solution that is 30% salt and 70% water. He has a solution that is 20% salt and one that is 50% salt. How much of each should he use to make the desired solution?

17. In two years, Kara will be three times as old as Leslie. One year ago, Leslie was one-sixth the age of Kara. How old are they now?

18. Dustin's boat traveled 36 miles downstream in three hours. The same boat traveled 30 miles upstream in five hours. What is the speed of the boat and the speed of the current?

19. Find the solution that will satisfy all three equations.

   A. $2X - 3Y + 3Z = 9$
   B. $4X + Y - 2Z = 0$
   C. $-6X - 2Y + Z = 0$

20. Name the two components of a vector.

FINAL EXAM

Simplify or put in standard form.

1. $(X^7 \div X^3) + (X^2 \cdot X^2) =$

2. $\dfrac{A^5 B^{-3}}{B^3 A^2} =$

3. $\left(\dfrac{8}{27}\right)^{-\frac{1}{3}} =$

4. $2\sqrt{5} + 7\sqrt{5} =$

5. $\dfrac{X}{3+i} =$

6. $\dfrac{3}{1+\sqrt{3}} =$

Add or Subtract.

7. $\dfrac{5}{6X} + \dfrac{4}{3Y} =$

8. $5Q^{-1}RQ^2 + 3QR - R =$

Solve using scientific notation.

9. $(.0009)(.027) =$

10. $\dfrac{3,700,000}{.002} =$

Solve for the unknown.

11. $2X^2 - 9X = 35$

12. $X^2 + 4X - 4 = -3X$

Find the solutions for each pair of equations. Sketch a graph of each equation, and show the solutions.

13. $Y = X^2 + 2$
$Y = X + 2$

14. $X^2 + Y^2 = 1$
$X^2 - Y^2 = 1$

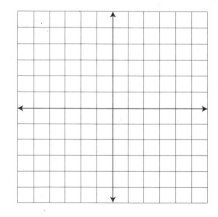
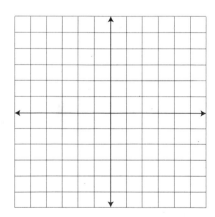

Answer the questions.

15. A new computer is being discounted 15%. If the original price was $1,565, what is the new price?

16. The atomic weight of sodium (Na) is 23 and that of chlorine (Cl) is 35. What is the percentage of sodium in NaCl?

17. The ratio of cats to dogs is 5 to 18. If there are 10 cats, how many dogs are there?

18. There are .62 miles in 1 kilometer. How many miles are there in 10 kilometers?

19. Michael and Alexandra left their home at 8:00 AM to drive to New York. Michael drove at 55 mph and arrived at 5:00 p.m. Alexandra drove at 45 mph and arrived at the same place as Michael. What time did Alexandra arrive?

20. I have 15 coins in my pocket. They are all either dimes or quarters. The value of the coins is $3.15. How many of each coin do I have?

21. Find three consecutive even integers such that three times the first, plus two times the second, minus the third equals 16.

22. A landscaper wants 100 pounds of grass-seed mixture that is 45% of type A seed and 55% of type B. He has a mixture that is 10% type A and one that is 60% type A. How much of each should he use to make the desired mixture?

23. In six years, Rose will be two times as old as Anne. Four years ago, Anne was one-third the age of Rose. How old are they now?

24. A boat can go 26 miles downstream in the same time it takes to go 6 miles upstream. The rate of the water is 5 mph. What is the rate of the boat?